The STEM Dilemma
Skills That Matter to Regions

Fran Stewart

2017

W.E. Upjohn Institute for Employment Research
Kalamazoo, Michigan

Library of Congress Cataloging-in-Publication Data

Names: Stewart, Fran, author.
Title: The STEM dilemma : skills that matter to regions / Fran Stewart.
Description: Kalamazoo, Michigan : W.E. Upjohn Institute for Employment
  Research, 2017. | Includes bibliographical references and index.
Identifiers: LCCN 2017044691 | ISBN 9780880996402 (hardcover : alk. paper) |
  ISBN 0880996404 (hardcover : alk. paper) | ISBN 9780880996396 (pbk. : alk.
  paper) | ISBN 0880996390 (pbk. : alk. paper) | ISBN 9780880996419 (ebook)
  | ISBN 0880996412 (ebook)
Subjects: LCSH: Science—Study and teaching—Economic aspects. |
  Technology—Study and teaching—Economic aspects. | Engineering—Study and
  teaching—Economic aspects. | Mathematics—Study and teaching—Economic
  aspects.
Classification: LCC Q183.3.A1 S75 2017 | DDC 331.11/4—dc23 LC record
  available at https://lccn.loc.gov/2017044691

© 2017
W.E. Upjohn Institute for Employment Research
300 S. Westnedge Avenue
Kalamazoo, Michigan 49007-4686

The facts presented in this study and the observations and viewpoints expressed are the sole responsibility of the author. They do not necessarily represent positions of the W.E. Upjohn Institute for Employment Research.

Cover design by Carol A. S. Derks.
Index prepared by Diane Worden.
Printed in the United States of America.
Printed on recycled paper.

To Eliot, Ethan, and the next generation of workers

Contents

Figures

Tables

Acknowledgments

This book—and the dissertation process that gave rise to it—would not have been possible without the help and guidance of many who deserve my thanks.

My committee members—Edward W. (Ned) Hill, Nicholas C. Zingale, and Wendy C. Regoeczi—provided insightful comments, meaningful guidance, statistical tutorials as needed, and ongoing support. Ned's friendship, recommendations, and sharp edits were indispensable in turning a long academic endeavor into a manuscript of relevance to policymakers, practitioners, and scholars.

My sister, Alice C. Stewart, was instrumental in starting me down this path and served as sage adviser, coach, and cheerleader all along the way. Her intellectual input, practical experience, and sisterly love were—and are—invaluable.

Ziona Austrian, Ellen Cyran, and the Center for Economic Development within Cleveland State University's Levin College of Urban Affairs allowed me access to and provided me help with some of the data.

I am grateful to Kevin Hollenbeck, Upjohn's former director of publications, for providing me an opportunity to share my work and findings with a wider audience. His insights and suggestions, as well as those of the two anonymous reviewers, helped to strengthen the research and crystallize the message. I also appreciate the contributions of Bob Wathen in editing, clarifying, and improving the experience for readers.

Lastly, I want to thank my family—David, Eliot, and Ethan—for their love and support throughout this years-long process.

1
More STEM Degrees

Economic Best Bets for Regions?

Within two months of taking office in December 2015, Kentucky Gov. Matt Bevin courted controversy with, in essence, an academic winners and losers list of public support for higher education: electrical engineers, yes; French literature majors, no. He articulated a simple policy reorientation toward incentivizing "things people want" (Beam 2016). Market forces have been reshaping college-going for decades, as individuals increasingly view further schooling in terms of professional development a necessity rather than a personal development luxury, but Bevin's proposal made plain public education's fundamental role as a tool of workforce and economic development.

Criticism of the proposal came swiftly from humanities professors, university presidents, and advocacy groups, as well as from more than a few political leaders in the state who no doubt had qualms about reinforcing Kentucky's poor reputation regarding education, including school funding and literacy rates that repeatedly rank near the bottom among all states (Ellis 2011). Yet, Bevin's vision of allocating public subsidy based on perceived workforce demand and best bets for investment return is not far outside the mainstream. Facing intense budget pressures following the Great Recession, 32 states had adopted some level of performance-based funding for two-year and four-year public postsecondary educational institutions, and another five states had such plans in the works in 2015 (National Conference of State Legislatures 2015). Many of these outcome-oriented measures include targets for degrees or certifications in science, technology, engineering, and mathematics (STEM) fields. In addition, a number of states use grant dollars to entice more high school graduates to pursue STEM majors in college.

ALL IN ON STEM

The rising preeminence of STEM is remaking primary and secondary educational systems as well, as evidenced by the renaming of local schools, realignment of curricula, and growth in after-school programs and ancillary activities. Examples include Science Club in Chicago, California Tinkering Afterschool Network, GirlsStart in Austin, Texas, and SHINE, which provides mostly low-income students in rural Pennsylvania exposure to STEM curricula.

Pursuit of STEM degrees has moved from one of personal interest or professional ambition to a matter of economic imperative and public priority. The policy assumption is clear: economies benefit from more scientists making discoveries, more engineers solving problems, and more computer experts programming solutions. In fact, President Obama issued an "all-hands-on-deck" alert for the critical mission of encouraging more students to pursue STEM degrees, setting a 10-year goal of 1 million more college graduates in such fields by 2020 (Office of the President 2012).

Despite the certainty evinced by the widespread federal, state, and local policies aimed at growing STEM education, there is only limited research demonstrating that a greater supply of STEM-degreed workers actually brings about the expected public gains. Certainly, there are reasons to assume such benefits will accrue: workers in various science and engineering occupations tend to earn wages that considerably exceed the average. Computer technology has transformed work environments and home life, creating demand for related skills. Areas home to products and activities emerging out of technological innovations have experienced the advantages of growth-stage industries.

Yet, the broad support for policies to increase the supply of STEM degrees obscures complexities and disregards contradictions. This is especially true when national goals and gains are adopted and applied locally. Does a larger share of STEM-degreed workers really improve the regional economy? Do regions have similar demand for such talent? Does promotion of STEM degrees—or degrees in general—neglect other avenues for workforce investment? What are the human capital best bets that can be made to address regional workforce challenges, align with opportunities, and advance regional economic well-being?

BEYOND DEGREES: BROADENING THE VIEW
OF REGIONAL HUMAN CAPITAL

So many policies and programs aimed at increasing the supply of twenty-first century talent come with a terse one-size-fits-all tag—more college is good, and more STEM is even better. STEM degrees have come to dominate the discussion regarding critical workforce needs, crowding out other paths toward acquiring knowledge and skill, and overshadowing other investments in human capital.

We are in a period when technology has broadened the reach of firms and individuals to engage in world markets. Ideas and information are exchanged virtually instantaneously, answers to even the most random questions are readily available, and change happens at a rapid pace. Yet the path to developing the knowledge and skills needed to operate within this environment has narrowed to the pursuit of one of a handful of educational degrees that take years to achieve.

This book details research directly inhabiting the muddled space where education policies and economic goals intermingle. The research expands beyond the current preoccupation with STEM degrees to explore the knowledge and skills occupations require. It sheds light on fundamental questions underlying Gov. Bevin's proposal and the actions of so many other political leaders: What is the appropriate role for public investment in knowledge and skills? What human capital development yields the greatest public return? Do these roles and forms vary by place?

Four assumptions lie at the heart of this research and challenge the orthodoxy of current STEM-oriented workforce development approaches to higher education and job training:

1) Differences in human capital deployment are key to understanding regional competitive advantage and economic well-being.

2) Jobs are a mix of specific knowledge and skills, including STEM-related skills but also generic skills, such as problem solving and communication. Generic skills are transferable and foundational and are commonly described as "soft."

3) Regions have distinct demands for different combinations of human capital and are the geographic units that best represent the function of labor markets.

4) Unique regional human capital is best represented not by the postsecondary degree attainment of its population but by the knowledge and skills required by its mix of jobs.

Research findings do support efforts to shift public resources toward the specific development of valuable technical skills. However, findings also indicate misconceptions and misperceptions about human capital concentrations and contributions, suggesting that the connection between human capital and economic well-being is not nearly as straightforward as STEM advocates suggest or as most STEM policies presume.

A COMPLEX RELATIONSHIP: HUMAN CAPITAL AND REGIONAL ECONOMIC WELL-BEING

Regional economic well-being is a multifaceted concept and requires a broader measurement than the commonly accepted focus on wages. Markers of healthy regions are higher wages, vigorous growth in gross regional product, higher productivity, higher per capita incomes, and lower poverty rates. Public policymakers often assume that human capital investments have across-the-board beneficial effects, but this study reveals a more complicated reality. Generic human capital investments impact some measures favorably, whereas other measures remain unchanged or even worsen.

Regions with greater human capital development do, by and large, enjoy higher wages. Yet, the effects of human capital development on regional output, productivity, per capita income, and poverty are much less straightforward. There is support in this work for the importance of higher STEM skills to regional economic well-being, but not for the rather narrow view of policies that strongly favor or promote increasing the number of scientists and engineers.

This study makes use of existing federal data collection of occupational knowledge, skill, and ability requirements and regional occupational employment levels to delineate differences in human capital needs and concentrations. Through a more fine-grained approach to operationalizing the concept of human capital than is found in discussions of STEM occupations, a new and more complete picture of how human capital impacts regional economic well-being emerges. Some of these new insights open new avenues for policy. This research exposes the thorny challenges associated with developing an economically viable workforce. Key conclusions from the research include:

- Degrees matter more to individuals than to the regional economy. Thus, policies that narrowly focus on college degrees may not be what benefits regions the most. Occupational knowledge and skill requirements are better measures of regional human capital than the commonly used proxy of degree attainment.

- STEM is more than scientists and engineers. Occupations requiring higher-than-average STEM skills are important to regional economic performance, but such occupations may not require a college degree.

- Not all high-paying jobs require STEM degrees or skills. Occupations with higher STEM requirements do tend to pay higher wages, but *so do* occupations demanding higher "soft" skills (i.e., critical thinking, teamwork, and communication). Occupations that pay the *highest wages* are those requiring both higher STEM and higher soft skills.

- STEM investment isn't necessarily a jobs program. Occupations with *higher* STEM requirements tend to employ disproportionately *fewer* workers.

- Low-skill, low-wage jobs predominate in most regions. Despite the policy focus on growing the supply of workers to fill "high-skill" jobs, *more than half* of all U.S. employment is relatively *low skill*. Large concentrations of low-skill employment drag down regional economic well-being.

- Regional differences in demand matter. The region with the largest share of employment accounted for by engineers, scientists, software developers, and similar STEM occupations had five times more STEM employment than the region with the smallest share of such occupations. Some regions have nearly 60 percent of their employment in occupations requiring a bachelor's degree, whereas other regions have 60 percent or more of their employment in low-skill occupations.

- Regional economic development requires more than just STEM workers, or even investments in human capital. Regional human capital is important, but it can only partly explain why some regions perform better than others.

STEM: A TRICKY PUBLIC INVESTMENT

Implicit in many STEM initiatives is the belief that a larger pool of workers educated in STEM will lead to the technological innovations, new products, and new processes that drive employment growth and economic well-being. Economists have long refuted the notion that increased supply of a product creates increased demand for that product (this is a refutation of what is termed Say's Law). No doubt, workers with understanding and mastery of modern technologies will be more likely to build on and expand new technologies. Yet, is mastery of specific technical skills what generates new products and markets or is it an entrepreneurial talent for observing the environment, envisioning opportunities, sizing up risk, and persisting in the face of obstacles and failures? Focusing too sharply on the technical aspects of innovation minimizes the importance of other knowledge, skills, and abilities, such as problem solving, critical thinking, teamwork, communication, and personal resilience.

Place-based initiatives that aim to grow the supply of STEM workers as a tool of economic development also run the risk that the newly developed human capital investments will not remain rooted in place. This newly minted supply can easily migrate to where demand is stronger and earnings are higher, or where tacit knowledge related to the occupation is being created. Well-educated, young workers tend to be highly mobile in any case, meaning they are likely to take their in-demand skills with them if there is not some rewarding job, emotional attachment, or area amenity holding them in place. A 2016 examination of American Community Survey and Census data by the *New York*

Times Upshot (Bui 2016) reveals the winners and losers in attracting educated talent. In general, college graduates have been leaving the Midwest and Northeast in favor of the West Coast and South. From 2000 to 2015, states such as Ohio and Michigan lost 4 percent or more of their college-educated workers age 40 and younger. Many of these migrating graduates owe at least part of their college educations to the states they ultimately fled. In essence, these states have become human capital exporters.

THE "ME TOO" TRAP

Many of the STEM initiatives adopted by state and local governments have little regard for the differences of place. This approach drives efforts to mimic the skill mix of Silicon Valley and other thriving technology hubs. Although this may be a winning strategy for areas that have the right conditions for STEM employment, not all places are competitive imitations of Silicon Valley, with rich veins of venture capital, large benches of experienced managers of high-growth firms, and—most importantly—a deep pool of talent and institutional support aligned to the high-technology product structure. Areas lacking in such locational advantages are likely to see their efforts at using the public policy equivalent of spontaneous generation either wither or fail to take root.

Areas are bound by their own industrial heritages and, to a certain extent, are built and buffeted by forces beyond their control. A legacy of dominant industries may leave some areas pockmarked with deep pools of obsolete talent when such industries decline or retreat, whereas other areas get lucky. What would the modern tech hubs of San Jose and Seattle be if William Shockley hadn't chosen to launch his nascent silicon semiconductor business near his ailing mother or Bill Gates hadn't relocated Microsoft closer to his boyhood home? Initiatives that assume investments in a larger pool of STEM talent will catalyze sustained growth through radical innovation and disruption ignore the tendency of innovation to be incremental in nature, building on *existing* platforms and strengths.

The business strategy literature acknowledges vagaries of external forces beyond the ability of managers and firms to predict. Yet, it makes clear that competitive advantages emerge out of how managers and firms *uniquely* respond to these external forces. This is the critical element in creating value and sustaining competitive advantage—not the sameness of development, not retracing trajectories of other firms, but deploying unique resources toward opportunities that leverage existing strengths, minimize risk, and align with market demand. A lesson to be transferred from business strategy to economic development strategy is that economic development policy must be attuned to the specific strengths and resources of a specific location. Strengths are often embedded in past industrial investments; thus, an understanding of how traditional industries are evolving is key. In addition, as local industries naturally evolve into new versions of themselves, workforce development must keep pace.

REGIONAL INVESTMENT IN HUMAN CAPITAL DEVELOPMENT: A DELICATE BALANCE

How human capital concentrations relate to regional economic well-being is highly dependent on product cycles and industry performance. The findings presented here offer strong support for an integrated view that "what regions do," to quote Feser (2003), is due in large part to what regions make now and made in the past. The ways in which a region's economic history can shape its future suggest a need for better alignment of human-capital-based policies with industry needs and expected performance. This would enable the prioritization of immediate talent demands as well as the identification of skill sets on which to build for the future. After all, most regional economic development is evolutionary, not revolutionary, and most technology-based development is found when new technology is pulled into existing products, rather than in new products that were pushed out from new technologies.

Aligning human capital investments too closely to existing industry needs does present the risk of an over-supply of workers in what may become legacy occupations. Over-supply, or redundant supply, of people in legacy occupations is a problem that many older industrial regions have faced when the skills of their workforce were tied too tightly to dominant industries that fell into decline. Although there is a risk associated with inaction, action based on faulty assumptions is risky as well. Regions that focus primarily on increasing the supply of workers with college degrees without regard to the local demand for human capital may find themselves with well-educated workers who are underemployed. These regions may see their investments depart for other areas where job prospects match the workers' newly developed skills. Either way, the regions' investments fail to achieve the desired results.

In some real sense, regional investments in workforce development involve three sets of comparisons: 1) weighing the value of skills with broad application against those with more narrow importance and worth, 2) determining support for skill sets with immediate relevance versus those that may bring future gains, and 3) recognizing that skills with broad benefit to the region may not be as rewarding to individuals. Evaluating trade-offs across these three dimensions represents a delicate balance for individuals, employers, and government-supported providers of education and training. The task of crafting regional human capital strategies is made more difficult by the fact that regions, despite being fundamental economic units, are rarely polities. Instead, regions encompass a number of political jurisdictions, all acting on their own set of needs with little effort spent on collaboration for the good of the region overall. In addition, funding for investments in regional human capital often comes through state and federal agencies, reflecting their goals and priorities, which may or may not match regional ones.

WHAT TO DO? RESHAPING POLICIES FOR REGIONAL ADVANTAGE

This research reveals insights for the Kentucky governor and other policymakers who want to pursue human capital development as a path to better economic well-being. Although Gov. Bevin's proposed plan to incentivize "useful" majors and other plans like it are made at the state level, they impact regional workforces. Findings from this research indicate several ways in which state and local policymakers and practitioners could refine and recalibrate their efforts directed at developing knowledge and skills critical to current and future advantage.

- Let data drive decisions (or at least inform them). Choices about investments in human capital development are often shaped by perceptions and observations that may or may not reflect ground-level realities for individual regions. Analyses of primary industries, specific occupations, and associated skill sets would provide the type of region-specific contours of human capital deployment that enable better targeted interventions.

- Cultivate collaboration. Regions may best represent the functioning of distinct labor markets, but policies that support them tend to be an amalgam of local actions and state and national initiatives. Given that few regions have political jurisdictions that mirror their geographic boundaries, the challenge falls to economic developers, regional education and workforce training providers, local mayors, city managers, and council members to build relationships, identify strategies, and encourage cooperation toward mutual place-based benefit. In addition, a collective regional voice serves to amplify the message that state and national skill-development strategies should recognize and be attuned to regional differences in demand.

- Avoid imitation. Resist the urge to think of human capital needs as uniform and to develop imitative policies. It's important to study best practices for insights into what works—at least, what seems to be working in certain regions. There will be shared goals and broad strategies, but the workforce needs of regions are highly individualized. Opportunities for competitive advantage arise from difference, not from sameness. Being in a position to seize on opportunities that arise from differences highlights the importance of a thorough and periodically updated understanding of immediate and near-term needs that are shared within industries or cut across multiple industries.

- Learn from the regional experts—employers—but don't let anecdotes drive decisions. Business leaders and industry advocates should be welcome partners, shaping understanding of immediate and projected human capital needs, but business reluctance to train and invest in developing its workers does not necessitate government intervention.

- Focus on fungible skills. The role for public support, especially at the regional level, is in developing fungible capacity, meaning skills and knowledge bases shared by entire industries or those that cut across multiple industries. These skills serve as the connective tissue and building blocks of a dynamic, adaptive workforce. Developing very job-specific skills should be left to employers.

- Explore many paths to important skills. Critical-thinking skills are considered key by-products of STEM education. But other higher education pursuits—business strategy courses, communication classes, and, yes, liberal-arts studies—develop critical-thinking skills in students. The challenge is having institutions demonstrate that they inculcate valuable and largely fungible cognitive skills that develop in concert with specific knowledge domains. It is also important to recognize that the same technological forces reshaping other industries will disrupt and remake learning models, as well. Online education has grown both as a disruptive enabler of distance learning and as an expander of traditional classroom activities. However, video games, virtual reality programming, and one-on-one mentoring, in person or assisted by technology, all have the potential to augment training and skill development outside formal educational and workplace settings.

- Balance immediate needs versus important future demands. The appropriate role for national and state governments may be to encourage high aspirations for human capital development and to shape a view of the future workforce. As the fundamental economic unit, however, regions must keep an eye on immediate needs while also strategically envisioning and assessing opportunities.

- Connect the next generation of workers to the work of the region. Develop programs or support opportunities for internships, mentoring, and apprenticeships. Opportunities for hands-on learning and career exposure should particularly be directed at the high school level. Make sure that high school students receive career guidance, and support interactions between high school guidance counselors, teachers, and local employers, especially manufacturers. Recognize that skill development and career preparation begin early. The math and science classes that students take in high school flow out of actions and decisions made in intermediate grades. Early exposure to work opportunities is critical to shaping learning choices and career aspirations.

- Pay attention to the bottom of the skills spectrum. Although so much of the focus of human capital interventions is on growing the share of workers with a high level of knowledge and skill, occupations requiring very little of their workers make up a far greater share of employment. As this research makes plain, larger concentrations of low-skill employment represent a significant drag on regional economic well-being, offsetting or even exceeding the gains from higher skills. Low-skill jobs provide critical access to the labor market, but the associated low wages and relatively limited opportunities for advancement present challenges for workers and regions alike.

- Brace for change and disruption. Regions know all too well how advancements in technology and changes in industrial processes have reshaped workforces. Industries that once employed thousands now produce more with only a handful of workers overseeing automated systems. Industries that once dominated local economies have shrunk, and even greater disruptions loom: self-driving vehicles, workerless stores, and self-directed learning, among others, will transform regional skill demand while also transforming the very nature of work.

MAKING THE CASE: EXPLORING AN OCCUPATION-BASED VIEW OF REGIONAL HUMAN CAPITAL

The remainder of this book presents a method for exploring regional human capital based on the knowledge and skill requirements evident in each region's portfolio of occupations. Chapter 2 provides a brief overview of key insights drawn from theory and highlights the ways in which these insights are being distorted in practice. Chapter 3 details the use of existing federal data to categorize occupations by the intensity of their STEM skill requirements, as well as their demand for "soft" skills. Chapter 4 examines the relationships between these categories, which reflect occupational skill requirements and wages. Chapter 5 applies the occupational skill requirements to regional employment levels to explore the effects of variation in human capital concentrations on regional economic well-being. Chapter 6 tests the explanatory power of regional human capital concentrations based on these occupational skill requirements within different educational attainment and population contexts. Chapter 7 offers an alternate method of categorizing occupational requirements to explore the effects of "middle skills" (the middle third of the distribution of skill levels) on regional economic well-being. Chapter 8 lists skill requirements shared across the occupational categories. Chapter 9 suggests ways in which considering variation in regional human capital deployment, manifest in occupational skill requirements, can be used to shape policy and practice.

2
Misunderstandings and Misapplications

Anybody who fires up a laptop to Skype with a telecommuting family member or spends a workday in a sparsely populated factory monitoring the activities of robots understands the impact of technological progress. Computers, automation, and the digitization of increasingly complex processes have disrupted and transformed work life and home life in ways only science fiction writers envisioned a half century ago.

Those with the entrepreneurial spirit and savvy to seize on advancing technologies have become titans of commerce, supplanting the industrial powerhouses of the mass production era. Workers with the combination of talent and good fortune to find a place within companies that benefit from disruptive technologies have also enjoyed the benefits of higher wages and innovative work environments. With the knowledge and skills needed to take advantage of technological change and create new products and services, computer programmers, mathematicians, engineers, and scientists have become workforce rock stars.

Political leaders, policymakers, and media pundits are, by nature, keen observers of their surroundings and have come to embrace technology education as *the* pathway to a secure economic future. They have responded to the mushrooming of computers and technology in the workplace with clarion calls for workers with more advanced skills. In addition, vocal business leaders, advocacy groups, and a spate of reports benchmarking gaps and lagging indicators have brought the technologically driven challenges facing the current and future workforce into sharp relief.

It's a short leap from observing the preeminence of technology in the modern economy to enacting policies that codify the supremacy of certain knowledge and skills. Hence, we have seen the rise in rather parsimonious funding formulas that support STEM education while minimizing the importance of other knowledge domains and skill pathways, all in the name of economic development and growth.

Yet, there are many reasons to question whether the line between STEM degrees and economic well-being is as straight as presumed, especially at the regional level.

SIX STEM FALLACIES

Alternative Realities: Theory Differs from Practice

The theoretical underpinnings of so many of today's policies promoting STEM education were laid a half century ago, when management guru Peter Drucker (1969) elevated the value of the "knowledge worker" and economist Gary Becker wrote the book *Human Capital*. Becker (1964/1993) formalized the long-observed connection between superior skill and higher wages, demonstrating that individuals' choices about developing their own capabilities had private and public economic consequences. Twenty-five years later, economist Paul Romer (1990) elevated the technology and innovations emanating out of expansions of knowledge and human capital to critical components of sustained economic growth.

A large body of literature supports the connection between greater skill and higher wages, greater productivity, greater economic output, and all manner of positive outcomes, ranging from better health to increased political engagement. Yet, a number of studies have failed to find the assumed benefits. Caselli, Esquivel, and Lefort (1996) found little support for economic growth from increasing levels of human capital. In exploring worldwide growth in schooling, Benhabib and Spiegel (1994) and Islam (1995) observed negative economic returns to human capital accumulation in some countries.

Moreover, the literature suggests an ambiguous connection between human capital accumulation and growth in employment (Bartik 1992; Cooper, Gimeno-Gascon, and Woo 1994; Holzer and Lerman 2007; Lerman 2008; Scott and Mantegna 2009; Shapiro 2006). The technological change that emanates from human capital investment and drives economic growth frequently leads to labor-saving devices and automations that remake work environments and eliminate jobs (Autor, Levy, and Murnane 2002, 2003).

Fluid Subsidies: Regions Are Not Mini-Nations

Regional human-capital-based policies designed to increase educational attainment broadly, and STEM education specifically, often are shaped by and receive some support from federal programs. But regional borders are far more porous than national ones. This means that human capital investments funded by regional or even state sources come with the very real risk that well-educated, or newly trained, workers will migrate to other areas for higher wages and better job opportunities, undercutting or negating any expected return on the investment of public resources. Certainly, failing to invest in human capital is a risk to regions; the literature indicates that better-educated regions tend to have higher productivity, higher wages, greater growth in per capita incomes, and higher job growth (Glaeser and Saiz 2003; Gottlieb and Fogarty 2003; Rauch 1993; Wolf-Powers 2013). At the same time, regions that have grown their share of the population with at least a bachelor's degree have experienced mixed results in terms of economic performance and public benefit (Andreason 2015).

The Department of Computer Science at the University of Illinois at Urbana-Champaign is a good case in point. The department is unquestionably excellent. It was one of the first of its kind when it evolved out of Illinois' Digital Computer Laboratory in 1964 and added graduate degree programs two years later (University of Illinois Department of Computer Science, n.d.-a). *U.S. News & World Report* has recognized the department's undergraduate and graduate programs as fifth best in the nation, and the average starting salary for graduates who completed a bachelor's degree in the 2014–2015 academic year was $85,027 (University of Illinois Department of Computer Science, n.d.-b). Graduates are associated with Mosaic—the Web browser that helped to popularize the Internet—PayPal, and YouTube (University of Illinois Department of Computer Science, n.d.-a). These are results that make economic developers beam, except that Netscape (the successor to Mosaic), PayPal, and YouTube are all located in Silicon Valley. Champaign-Urbana is a hotbed of world-class computer science; it is not a hotbed of digital startups.

Enrollment in the Department of Computer Science offers the prospect of an excellent return on a parent's tuition payments. However, its graduates are, more often than not, working outside the state, let alone the region. Therein lies the national versus local public policy challenge. Illinois' Department of Computer Science is a vital source of talent for Silicon Valley and a national treasure, but the return from subsidies provided by Illinois taxpayers is less clear. Illinois residents whose children join the annual spring migration to the Bay Area receive a return on their investment, if not the return of their children. And California's employers and taxpayers do not even bother to send thank-you notes for the subsidies the State of Illinois provided to their highly skilled immigrant workers.

Wrong by Degree: Education Fails to Capture the Breadth of Human Capital Theory

Human capital theory as put forth by Schultz (1961), Arrow (1962), and Becker (1962, 1964/1993) was a broad concept encompassing all manner of investments that enabled workers to be more productive, but practically speaking, educational attainment—the highest level of schooling completed—has come to serve as the standard proxy for human capital. Data on education are routinely collected and readily available. Ease of access and the ability to compare different levels of attainment and expenditure across nations and regions have imparted education with a practical relevance that other potential measures of human capital, such as training, experience, or self-study, may lack. The proxy measure for the broad concept of human capital has become its synonym—its de facto meaning. The concept has been narrowed to fit the variable used to measure it.

Formal educational attainment is but a blunt and imperfect operationalization of the broad concept of human capital. Having a particular degree or level of education is not necessarily the same as having competence in a particular set of knowledge, skills, and abilities. Degrees alone mask a wide range of economic return on relatively similar investments of money and time. For example, in an analysis of college majors, Carnevale, Cheah, and Hanson (2015) found that majors in top-paying fields returned $3.4 million more over a lifetime than the bottom-paying majors. Entry-level workers with STEM degrees had median wages of $38,000 in 2013 dollars, compared with $29,000 for workers with humanities degrees. Moreover, mid-career workers with STEM degrees earned 50 percent more in 2013 than mid-career workers with liberal-arts degrees (Carnevale, Cheah, and Hanson 2015).

Although a college degree typically imparts protection from unemployment, college graduates with a humanities degree in 2008 were far more likely to be without a job one year later than graduates with a business degree (13 and 9 percent, respectively) (Torpey 2013).

Despite the assumed connection between higher education and better job prospects, roughly 10 percent of recent college graduates were "idled" in 2016, meaning they were neither employed nor pursuing further education (Kroeger, Cooke, and Gould 2016). One out of every eight recent college graduates not pursuing further study was unable to find full-time work in 2016. Even more troubling, 44.6 percent of college graduates under the age of 27 were employed in jobs that did not require a college degree (Abel and Deitz 2016).

Say's Law Is Still Wrong: Human Capital Development Is Not the Same as Human Capital Deployment

Human-capital-based policies that are not anchored in the realities of regional labor markets are reincarnations of either Say's Law or *Field of Dreams*. This "build it and they will come" view assumes that a better supply of talent, especially STEM talent, will either attract firms that demand such talent or lead to the new products and firms that drive job growth, greater productivity, and higher wages.

But technological change doesn't "just happen." It is frequently incremental and complementary of existing technology and skills (Acemoglu 1998). In other words, context matters (Autor, Levy, and Murnane 2003). Regions tend to be locked into their industrial or occupational legacies, and change is largely incremental due to the reinforcing nature of path dependence (Martin and Sunley 2006; North 1990). This suggests that regional human capital policies are more likely to yield anticipated benefits if they are aligned to a region's specific labor demands and business strengths.

Trade Comes from Being Different: Differences in Human Capital Deployment Are Important Opportunities for Competitive Advantage

Regions (and states) are tending to enact rather similar, boilerplate human-capital-based policies, all with the premise that more college is good, and more STEM is even better. Business strategy literature, however, presents the view that human capital is valued because of its unique fit with the needs and opportunities of firms. Human capital that is rare, difficult to copy, and matched to the specific strengths of firms is the critical element of sustained competitive advantage (Barney 1991). In other words, leveraging *differences* in human capital—not simply imitating competitors—should be the goal of individual firms within a region, as well as the region itself.

However, it is also important to keep human-capital-based interventions in context. Workforce issues are only part of the mix of factors businesses consider when they decide where to locate or relocate. And, it is only one factor among many that predict business survival rates. The considerable advantage of a deep pool of talent helps to root business activity in place, but it may not be enough to prevent a firm from transplanting its operations due to other factors. Some of these factors, such as better weather and the synergies gained by being closer to a key supplier or customer, are beyond the control of policymakers. Others, such as the tax environment and infrastructure, are traditional policy focal points. Still others, such as a tradition of union activity, shape impressions that can undercut the human capital advantages of place. The rise of automotive industry hubs in southeastern states where no legacy of such skilled activities existed demonstrates that businesses continually weigh advantages of human capital within the context of other locational strengths and weaknesses.

Hidden in Plain Sight: Too Heavy a Focus on STEM May Mask Other Valuable Skills

The theorized connection between technology and economic growth may lead policymakers and researchers to discount the importance of other types of human capital for economic success. Fifty years ago, Nelson and Phelps (1966) foreshadowed the importance of having educated scientists to keep up with change, but they noted that it was equally important to have educated managers to seize on opportunities and make decisions. Business executives describe attributes such as communication, problem solving, social skills, courtesy, responsibility, teamwork, and flexibility as critical worker attributes in today's work environment (Robles 2012). Workers who are able to apply their knowledge and skills toward complementing existing skill sets and industrial demands, as well as supporting emerging ones, will be more productive and, thus, more valuable to both employers and the regional economy (Lerman 2008). Moreover, there is evidence that the very sweeping technological changes that have made STEM skills the focus of policy attention have served to increase the importance of "people skills"—that is, the ability to interact, communicate, care for, and motivate others (Borghans, ter Weel, and Weinberg 2014).

TAKING A DIFFERENT ROAD

These six challenges to the presumed link between STEM degrees and regional economic vitality suggest that a different approach to understanding and enhancing regional human capital is warranted— an approach that readjusts the focus from the education supplied to the skills demanded. The remainder of this book presents a method for exploring regional human capital based on the knowledge, skill, and ability requirements associated with each region's portfolio of occupations.

3

A Method for Bundling Occupational Skills

Regions differ in their mix of occupations, and occupations differ in their mix of required skills. This simple reality underscores the complexity of regional human capital deployment. Yet, many policies directed at human capital development, especially increasing the number of STEM graduates, assume a rather uniform demand or, at least, a similar capacity to create demand.

Regional human capital deployment would best be captured at the level of a region's collection of jobs. More specifically, thinking of jobs as a bundle of knowledge, skills, abilities, educational requirements, and experiences (Bacolod, Blum, and Strange 2010) would more closely align to the broad concept of human capital and would provide insight into each region's particular alchemy of attributes. Human capital required by actual jobs would best explain how each region's unique stock of human capital is deployed and valued in the larger economy. Estimates of the distribution of human capital required by a region's pool of jobs also would provide the most detailed insight into the human resources that form the basis of sustained competitive advantage for businesses and the regions where they are located (Barney 1991).

However, an analysis of each region's unique mix of job-level talent requirements would be an onerous undertaking for regional policymakers intent on economic development. An alternative exists in using a federal database that can approximate the deployment of regional human capital without having to delineate the knowledge, skills, and abilities required of each region's mix of jobs. The dataset, the Occupational Information Network (O*NET), enables an exploration of human capital that is more reflective of the broad definition of the concept but that, like measures of educational attainment, is also widely available and easily accessible. Although O*NET data are national level and thus cannot capture region-specific occupational differences, the database is both in-depth and regularly updated in its detailing of the individual skills, abilities, and knowledge areas required of occupations, as well as the most frequently required levels of educational attainment, experience, and on-the-job training.

O*NET's extensive occupational mapping allows for a finer grained understanding of human capital—the stock of knowledge, skills, and abilities—associated with economic gain, both for individual workers and for regions. The O*NET database has been used in economic development research to assess the benefit of occupations requiring high- and mid-level STEM knowledge to regional vitality (Rothwell 2013). Scott (2009) and Florida et al. (2012) used O*NET data to demonstrate an increase in occupations requiring cognitive skill and a decrease in employment requiring physical skill. Koo (2005) explored O*NET to show the importance of occupational clusters to regional economic performance. Yakusheva (2010) demonstrated that the college wage premium is a function of the goodness of fit between field of study and occupation, and Maxwell (2008) drew on the O*NET database to identify skills that command higher wages among less-educated workers. The O*NET database has also attracted the attention of researchers in the areas of psychology, human resources, career guidance, and family relations.

O*NET OVERVIEW

Sponsored by the U.S. Department of Labor's Employment and Training Administration, O*NET was developed to supplant the Dictionary of Occupational Titles (USDOL, n.d. Hereafter, O*NET). With a stated goal of serving as "the nation's primary source of occupational information" (O*NET Online Help Overview), the O*NET database has been regularly updated and expanded since 2003. This research draws on Version 19.0, which was released in July 2014. Version 19.0 provides a detailed mapping of 942 occupations and includes a comprehensive update of 126 of those occupations. The O*NET method has received endorsements from hundreds of industry organizations and associations. The endorsements reflect the success of O*NET's mission of presenting what amounts to an ever-evolving rendering of the U.S. labor market supported by a network of workers in participating establishments who both contribute to and draw from the database of occupational requirements and expectations (O*NET website).

The foundational framework for O*NET is its Content Model, described as a "theoretically and empirically sound" system for guiding the collection and integration of information to develop a deep understanding of each occupation's mix of attributes. The Content Model divides six major informational domains into worker-oriented and job-oriented characteristics, as well as cross-occupation and occupation-specific attributes. The six domains are worker characteristics, worker requirements, experience requirements, occupation-specific information, workforce characteristics, and occupational requirements.

O*NET DATA COLLECTION

The O*NET Data Collection Program surveys incumbent workers (typically a random sample of two to three dozen engaged in each occupation of interest) to gather information on the knowledge, skills, abilities; education, experience, and training requirements of their jobs; as well as their work styles, daily work activities, and interests. Occupational experts drawn from trade or industry associations are asked to complete questionnaires for occupations that pose difficulties in identifying incumbent workers.

Although the O*NET questionnaires collect information from representative workers regarding daily tasks, preferred work styles, and personal interests, this research focuses exclusively on the knowledge, skill, and ability attributes of each listed occupation. The decision to limit the focus was guided by the existing career advising and human resources literature, as well as general practice; job descriptions are often built—and job applicants evaluated—based on key knowledge, skill, and ability (KSA) requirements. Information on each occupation's average level of education, experience, and training—drawn from the worker requirements and experience requirements domains—was also incorporated into this analysis.

Surveyed workers are asked to rate each of 120 KSA attributes: 33 knowledge domains, 35 individual skills, and 52 abilities. There is a level of overlap, especially among the skill and ability attributes. In the

O*NET Content Model, skill is conceptualized as a developed capacity, whereas ability is more an innate characteristic. Worker skills can be thought of as being built on individual abilities. For example, mathematical reasoning ability underlies mathematical skill.

Each KSA attribute is assessed along two dimensions. Surveyed workers are first asked to assess the *importance* of a specific attribute to their job performance on a scale of 1 to 5, with 1 equaling "not important" and 5 being "extremely important." For KSAs that rate a 2 or higher, meaning the attribute is at least "somewhat important," surveyed workers are then asked to rate, on a scale of 1 to 7, the *level* of the attribute necessary to perform their job. Workers completing the questionnaire are provided attribute-specific anchors to guide their rating. For example, workers who indicate that oral comprehension is at least "somewhat important" are then asked what level of oral comprehension their job requires. A 2 indicates a level sufficient to "understand a television commercial," a 4 indicates a level equal to understanding "a coach's oral instructions for a sport," and a 6 is the level of oral comprehension necessary to "understand a lecture on advanced physics."

The difficulty in using scale, or ordinal, data, which asks respondents to locate their observation along a 1 to 5 (or 1 to 7) numerical scale is that the intervals between the numbers do not measure distances. Whether the distance between 1 and 2 is the same as the distance between 4 and 5 cannot be known. Each score is a subjective assignment guided by the survey's instructions. The result is that the scores provide valuable insights by capturing respondents' perceived ordering of the importance of specific attributes and of the relative level of mastery of an attribute required to successfully complete a specific job.

IDENTIFYING STEM AND SOFT KSA BUNDLES

A growing number of studies have set out to explore the heterogeneity of demand for human capital by exploring the bundle of knowledge, skills, abilities, and other attributes required within and across occupations. However, few attempts have been made in the literature to explore the effects of STEM skills, particularly on regional economic well-being. Carnevale, Cheah, and Hanson (2015) compared the economic value of college majors, bundling degrees into supergroups, such as STEM, business, and humanities, for an exploration of entry-level and mid-career wages. Carnevale, Smith, and Melton (2011) used O*NET data to explore STEM competencies and demand, providing insight into career pathways and talent pipelines. Teitelbaum (2014) challenged the "myth" that the nation is falling behind in educating STEM workers. Xue and Larson (2015) observed geographic differences in demand for workers with STEM degrees. Rothwell (2013) was an exception in analyzing STEM employment at the regional level and focusing primarily on occupations requiring less than a bachelor's degree to address a perceived "high STEM" bias.

Evidence in the business literature shows that the intense policy focus on STEM may be misplaced. Business executives often cite a need for workers who can think through problems and communicate effectively (Robles 2012). Robles concluded that employers place a higher value on soft skills (personal attributes and interpersonal characteristics) than on hard (technical and specific) skills, but soft skills are often ignored in university and training curricula and the academic literature. In a review of empirical work on communication skills, Brink and Costigan (2015) found listening to be a critical but often underappreciated ability. Borghans, ter Weel, and Weinberg (2014) demonstrated that the speed and breadth of technological and organizational change over the past few decades have made "people skills" increasingly important in the labor market, even though such skills are more likely to receive attention in the psychology literature than in the economics literature. Generic skills (i.e., communication and problem solving) and occupation-specific skills have been found to be as important as the technical and "academic skills" that receive overriding focus (Lerman 2008). In addition, Gibbons and Waldman (2004) highlighted the importance of task-specific skills to labor demand, particularly in terms of career ladders and upward mobility.

The O*NET Content Model sorts *abilities* into categories of cognitive, psychomotor, physical, and sensory. It divides *skills* into categories of basic, cross-functional, and technical. *Knowledge* is divided into 10 domains: business and management activities, manufacturing and production, engineering and technology, mathematics and science, health services, education and training, arts and humanities, law and public safety, communications, and transportation.

Extracting only those KSAs that O*NET defines as involving science, technology, engineering, mathematics, or medicine could be presumed to reveal the understanding and capabilities that employers, the popular media, and political leaders mean when they advocate for more or better "STEM skills." Based on O*NET definitions, 35 of the total 120 assessed KSAs can be classified as "STEM skills": 14 *skills*, ranging from the obvious (math and science) to the less so (quality control analysis and troubleshooting); 17 *knowledge domains* (including social sciences, which the National Science Foundation [NSF] counts among STEM college majors); and 4 *abilities* (all having to do with numeracy and spatial facility). Guided by Rothwell (2013), certain *mechanical skill* and *ability* attributes were included in the bundle of STEM KSAs.

This methodology draws on both O*NET's *importance* and *level* scores to arrive at a score reflecting the intensity of each occupation's requirement for each of the KSAs of interest. Using only one dimension of the occupational assessment (as was done in Maxwell 2008; Rothwell 2013; Scott 2009; Scott and Mantegna 2009; Yakusheva 2010) loses some of the detail in understanding variation in how knowledge, skills, and abilities are deployed throughout occupations. For this analysis, the O*NET importance score and level score for each occupational attribute were multiplied together (as demonstrated in Abel and Gabe 2008; Florida et al. 2012; Hadden, Kravets, and Muntaner 2004) to derive a single score reflecting the intensity of each KSA for each occupation. The highest mean *intensity* scores across all occupations for each of the STEM KSAs are found in the knowledge areas of mathematics and computers and technology, which no doubt reflects the ubiquity of computers in today's work world.

Removing the STEM KSAs, as well as those measures defined by O*NET as reflecting psychomotor, physical, and sensory capabilities, left a collection of understandings and capabilities that reasonably can be thought of as operationalizing what is meant by the rather nebulous concept of "soft skills." In this manner, 50 of the total 120 KSA variables were placed into a "soft skills" group: 19 *skills*, which encompass active listening as well as time management; 14 *knowledge domains*, including language and philosophy; and 17 *abilities*, such as oral expression and problem sensitivity. Oral comprehension and oral expression had the highest mean "soft skills" scores across all occupations, an observation that supports and perhaps informs repeated references in the business literature and media regarding the importance of "communication skills."

This residual group includes some KSAs that may be thought of as more specific, or "harder," than the relationship and cognitive abilities typically identified as "soft skills." For example, customer and personal service, English language, and education and training have the highest mean intensity (O*NET *interest* times *level*) scores across all occupations among the knowledge domains included in the soft skills group. Facilitating relationships and understanding are central to these domains. Although school curricula often interpret "communication skills" as written expression, mean intensity scores highlight the importance of listening and speaking skills. The business literature and educational policies tout the importance of thinking critically and solving problems, but occupational requirements indicate a demand for workers who are able to recognize problems, prioritize information, and make decisions, as well.

Table 3.1 provides a list of the 35 STEM and 50 Soft KSAs. As noted earlier, many of the social sciences are included in the list of STEM KSAs based on O*NET and NSF definitions. The list of Soft KSAs includes knowledge domains such as history and philosophy. Although these specific disciplines may fall outside the broad applicability typically associated with "soft skills," such domains tend to be classified as part of the humanities. Given that many of the Soft skills deal with human interactions, disciplines that focus on the study of human culture and conditions would seem to be acceptably labeled "soft." The limitations of two broad KSA dimensions and the decisions to include all skills and abilities not defined by O*NET as physical or psychomotor and to include all knowledge domains drove these groupings.

USING STEM AND SOFT KSA BUNDLES TO CATEGORIZE OCCUPATIONS

Whether in political speeches, the popular media, or human capital literature, desirable skill sets are those that are—or are assumed to be—"high," especially in terms of STEM skills (Rothwell 2013; Teitelbaum 2014). High skills, both STEM and non-STEM, are assumed

Table 3.1 Sorting O*NET Knowledge, Skill, and Ability Attributes (KSAs)

35 STEM KSAs	50 Soft KSAs
Skills	Skills
Mathematics	Active learning
Programming	Active listening
Science	Complex problem solving
Operations analysis	Coordination
Operation and control	Critical thinking
Operation monitoring	Instructing
Systems analysis	Judgment and decision making
Technology design	Learning strategies
Equipment maintenance	Management of personnel resources
Equipment selection	Monitoring
Installation	Negotiation
Troubleshooting	Persuasion
Quality control analysis	Reading comprehension
Repairing	Service orientation
Knowledge domains	Social perceptiveness
Biology	Speaking
Chemistry	Systems evaluation
Computers and electronics	Time management
Design	Writing
Engineering and technology	Knowledge domains
Mathematics	Administration and management
Physics	Clerical
Psychology	Communications and media
Sociology and anthropology	Customer and personal service
Geography	Economics and accounting
Medicine and dentistry	Education and training
Therapy and counseling	English language
Building and construction	Fine arts
Food production	Foreign language
Mechanical	History and archeology
Production and processing	Law and government
Telecommunications	Personnel and human resources
	Philosophy and theology
	Sales and marketing

Table 3.1 (continued)

35 STEM KSAs	50 Soft KSAs
Abilities	Abilities
Mathematical reasoning	Category flexibility
Number facility	Deductive reasoning
Spatial orientation	Flexibility of closure
Visualization	Fluency of ideas
	Inductive reasoning
	Information ordering
	Memorization
	Oral comprehension
	Oral expression
	Originality
	Perceptual speed
	Problem sensitivity
	Selective attention
	Speed of closure
	Time sharing
	Written comprehension
	Written expression

to be in greater demand by employers, return greater reward to individual workers, and create greater economic prosperity for cities, regions, and nations. "Low" skills, conversely, are presumed to be in need of upgrading to access the assumed-to-be in-demand high-skill jobs and bring economic benefit to individuals, firms, and regions.

This methodology attempts to explore the KSAs of occupations within this binary high–low structure. Later chapters will present alternate approaches to exploring occupational requirements, including an attempt to identify the importance of "middle" skills in today's economy. It should also be noted that, reflecting the common vernacular, the term *skill* is frequently used throughout this book to refer to the entire bundle of KSAs.

As noted previously, this methodology multiplies O*NET's *importance* and *level* scores for each KSA of interest to arrive at a score reflecting the intensity of each occupation's requirement for the individual STEM and Soft KSAs. Each of the 942 O*NET occupations was assessed as to whether its intensity score was above or below the mean for each of the 85 total KSAs making up the STEM and Soft bundles. Occupations with intensity scores that were above the mean score were labeled "high," and those with below-average needs were labeled "low." For example, the O*NET occupational code signifying chief executives had a mathematics knowledge score of 14.4, which is above the mean of 11.0 for all 942 occupations. Therefore, the occupation was designated as "high" for that STEM knowledge domain. Conversely, the chief executives occupation code revealed an intensity score of just 2.1 for science skill, earning it a "low" designation for that attribute.

Each occupation then was assessed as to whether its requirement across the entire bundle of 35 STEM KSAs was above the mean for all occupations. If the occupation was above the mean, it was designated as "High STEM"; if it was below the mean, it was labeled "Low STEM." This process was repeated for the bundle of 50 Soft KSAs to label each of the 942 occupations as either "High Soft" or "Low Soft." For example, the O*NET occupation code corresponding to chief executives was found to have below-average intensity requirements across all 35 STEM KSAs, earning it a "Low STEM" designation. But its above-average intensity scores on 47 of the 50 Soft KSAs indicated the chief executive occupation to be "High Soft."

Although STEM skills and soft skills are frequently discussed separately, this research recognizes that occupations requiring high science or math skills also may require advanced problem-solving and critical-thinking skills. As such, occupations were ultimately categorized based on the intensity of their requirements along both dimensions. Combining the STEM and Soft labels revealed that 29 percent of O*NET occupations (273) require both above-average STEM KSAs and above-average Soft KSAs; 20 percent (188) require High STEM but Low Soft KSAs; 21 percent (199) require Low STEM but High Soft KSAs; and 30 percent (282) require both below-average STEM and Soft KSAs.

LINKING OCCUPATIONAL SKILL SETS TO OCCUPATIONAL WAGE AND EMPLOYMENT DATA

Simply categorizing occupations based on their STEM and Soft intensity may help reframe a discussion that frequently addresses such demands in isolation. Yet, exploring whether an occupation-based operationalization and measure of human capital presents greater insight for policymakers requires a method for linking the four STEM/Soft skill categories to occupational wage and employment data. For this analysis, each O*NET occupation was matched to wage and employment data available from Occupational Employment Statistics (OES)[1], a federal-state collaboration between the U.S. Department of Labor's Bureau of Labor Statistics (BLS) and state workforce agencies that surveys 200,000 establishments every six months over a three-year cycle.

Given that both O*NET and OES are based on the BLS Standard Occupational Classification (SOC) system, linking the two databases was a relatively straightforward process. However, O*NET uses a more fine-grained occupational classification than that available for the wage and employment data from the OES. The difference in occupational detail across the two data sources led to a number of incomplete matches that needed to be addressed. In addition, a few occupations were missing relevant data on median wages or modal educational, experience, and training requirements.

Ultimately, the O*NET occupational data on KSA intensity, as well as typical education, experience, and training expectations, were matched to OES national wage and employment data released in 2014 for 764 occupations. Roughly 45.8 percent of these occupations (350) were categorized as requiring above-average STEM KSAs; 44.1 percent required above-average Soft skills. Examining occupations on both dimensions revealed that 23.8 percent required above-average STEM and above-average Soft skills; 22.0 percent required above-average STEM but below-average Soft KSAs; 20.3 percent required below-average STEM but above-average Soft skills; and 33.9 percent required both below-average STEM and Soft skills.

Table 3.2 provides examples of occupations that were sorted into each of the four categories. As can be seen, the High STEM/High Soft category captures many of the high-education occupations that political leaders, pundits, and educators refer to when they talk about the importance of STEM skills. Software developers, engineers, mathematicians, scientists, and medical doctors inhabit this category, which reflects the highest occupational skill intensity among the four categories. Some occupations not often associated with STEM, such as industrial production managers, emergency management directors, and first-line

Table 3.2 Examples of Occupations Sorted by STEM/Soft Categories

Low STEM/High Soft	High STEM/High Soft
Chief executives	Computer and info. systems managers
Advertising and promotions managers	Industrial production managers
	Computer systems analysts
Marketing managers	Information security analysts
Sales managers	Software developers, applications
Personal financial advisers	Web developers
Statisticians	All engineers
Economists	Scientists
Budget analysts	Doctors/pharmacists/dentists
Mental health counselors	Science teachers
Lawyers	Survey researchers
Teachers	Emergency management directors
Reporters and correspondents	

Low STEM/Low Soft	High STEM/Low Soft
Tax preparers	First-line supervisors of mechanics, installers, and repairers
Insurance appraisers	
Writers and authors	Computer programmers
Radio operators	Software developers, systems software
Athletes and sports competitors	computer occupations, all other
Models	Mathematical technicians
Occupational therapy aides	Mechanical drafters
Physical therapist assistants	Civil engineering technicians
Crossing guards	Electro-mechanical technicians
Home health aides	Magnetic resonance imaging techs
Telemarketers	Surgical technologists
Retail salespersons	Medical equipment preparers
Customer service representatives	Derrick operators, oil and gas
	Computer numerically controlled machine tool programmers, metal and plastic
	Roustabouts, oil and gas

supervisors of mechanics, installers, and repairers, are also captured in this category. Various technical workers, such as computer numerically controlled (CNC) machinists, electro-mechanical technicians, medical equipment preparers, and oil and gas workers, fall into the category of High STEM/Low Soft skills. This would seem to provide support for relatively recent efforts to expand the concept of "STEM jobs" to include technical and mechanical occupations, many of which may have relatively low educational requirements. Occupations that involve relating to others, such as teachers, counselors, and managers, dominate the Low STEM/High Soft category. Occupations in the Low STEM/ Low Soft category include telemarketers, customer service representatives, and retail salespeople, as well as many occupations in the health-care industry.

Note

1. BLS, n.d. -a. Hereafter, OES.

4

STEM Skills, Soft Skills, and Worker Wages

Exploring the human capital reflected in occupational knowledge, skill, and ability requirements offers theoretical and practical advantages over the common proxy of educational attainment.

1) It more closely resembles the broad view of human capital shared by Schultz (1961), Arrow (1962), and Becker (1962, 1964/1993). It acknowledges education as an important path to the development of knowledge, skills, and abilities but also makes room for experience, practice, self-study, and Arrow's (1962) concept of learning-by-doing.

2) It presents a more fine-grained understanding of human capital by revealing how various knowledge, skill, and ability attributes cut across occupations.

3) It shifts focus away from human capital supply toward human capital demand. As the business literature makes clear, the value of human capital arises from how it can be applied to advantage. In other words, development of human capital, whether through education, training, or experience, is only part of the story; the economic value of human capital extends from its deployment. In economic terms, wage rates are determined by the intersection of supply and demand, not supply on its own.

A measure of human capital rooted in occupational demands would, by itself, seem a warranted reframing of the supply-based practice of assessing educational attainment levels of individual workers or entire populations. However, the method presented in Chapter 3 seeks directly to explore human capital contributions within the current policy focus on STEM and expand the analysis to include "soft" skills. A critical assumption apparent in the method is that occupations consist of a mixture of highly specific technical skills and more generically applicable skills. Human capital theory provides an expected earnings outcome based on this bundling assumption: occupations with higher skill demands will pay higher wages than those with lower skill demands as compensation for the time and effort required to master the skills. For this analysis, high skill is defined as those occupations with above-average requirements.

Growth in programs and schools designed to develop STEM skills and support STEM jobs yields another expected outcome: occupations that require a high level of STEM skills will pay higher wages than occupations that do not. This is based on the widely held expectation that demand for these skills is exceptionally strong.

Another expected benefit from building the STEM/Soft KSA categories is an improvement in predictive ability over the commonly used human capital proxy, educational attainment.

These expectations lead to four testable hypotheses:

H1. Occupations requiring above-average STEM and above-average Soft KSAs pay higher wages than occupations requiring other skill combinations.

H2. Occupations requiring above-average STEM but below-average Soft KSAs pay higher wages than occupations with low STEM skill requirements but lower wages than occupations with the highest skill requirements.

H3. Occupations requiring below-average STEM KSAs but above-average Soft skills pay higher wages than occupations with the lowest skill requirements but lower wages than occupations with High STEM skill requirements.

H4. Occupations requiring below-average STEM and below-average Soft skills are hypothesized to pay less than occupations requiring higher levels of skill.

Figure 4.1 summarizes the hypothesized relationships between occupational skill sets and median wage.

Figure 4.1 Hypothesized Relationships between Occupational Skill Sets and Median Wage

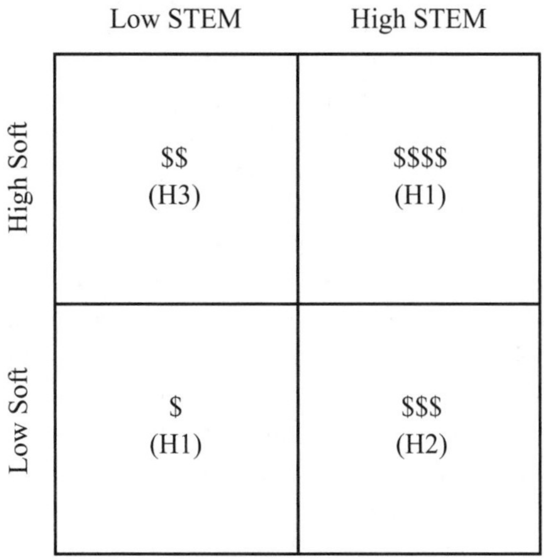

	Low STEM	High STEM
High Soft	$$ (H3)	$$$$ (H1)
Low Soft	$ (H1)	$$$ (H2)

METHODOLOGY

To test the hypotheses with regression analyses, the four STEM/ Soft human capital categories were recoded into three dichotomous variables, omitting the Low STEM/Low Soft category to serve as the reference group.

In addition to the STEM/Soft skill variables, data on education, experience, and training were extracted from the O*NET database to serve as control variables. O*NET's 1–12 education coding scheme, which ranged from "less than high school" to "post-doctoral," was recoded into a dummy variable with 1 indicating that the occupation required a bachelor's degree or higher and 0 indicating less than a bachelor's degree. The 1–12 coding scheme for experience was recoded into a dummy variable with 1 indicating more than a year of experience required and 0 indicating a year or less. O*NET's 1–9 coding scheme for on-the-job training was recoded into a dummy variable with 1 indicating more than three months of training required and 0 indicating three months or less.

Beyond these human-capital-related variables, a fourth control variable—occupational employment—was included in the model to address wide differences in the size of the occupational groups. The variable is the absolute number of jobs in an occupation. It was included to control for size, or scale, effects. Many of the very large occupational groups have low STEM scores and low Soft scores. For example, the 10 largest occupations in the Low STEM/Low Soft category alone account for nearly 21 percent of total U.S. employment. These occupations include retail salespersons, cashiers, customer service representatives, and waiters and waitresses. Although employment in these low-skill occupations tends to be large, the largest job losses related to the Great Recession were among low-skill workers, and recovery for such jobs has been sluggish. Carnevale, Jayasundera, and Gulish (2016) found a net 5.5 million fewer jobs for workers with a high school diploma or less, compared to 2007 employment levels. This would be expected to create downward wage pressure on these low-skilled occupations. The size of these occupations means that they may be exerting an effect on the labor market that is independent of skill alone.

The human capital literature frequently uses median wage to measure the return on some sort of human capital investment (e.g., Carnevale, Smith, and Strohl 2010; Feser 2003; Feser and Bergman 2000; Florida et al. 2012). This model follows suit.

Table 4.1 summarizes the independent, dependent, and control variables used in the regression analysis; both the wage and the employment variables were natural log transformed to help normalize their distributions.

RESULTS AND DISCUSSION

Table 4.2 provides frequency statistics for the KSA variables of interest, as well as the number of occupations requiring a bachelor's degree or higher and the number of occupations in the two experience and two training categories.

Table 4.1 How Variables Were Defined and Calculated for the Occupational Human Capital Analysis

Variable	Definition	Source
Dependent variable		
Median wage	Natural log of occupational median wage, transformed to adhere to regression assumptions about normalized distribution.	OES, May 2014
Independent variables		
High STEM/ High Soft	Dummy variable where occupations requiring High STEM skills and High Soft skills = 1; any other skill combination = 0.	Calculated using O*NET 19.0
High STEM/ Low Soft	Dummy variable where occupations requiring High STEM skills but Low Soft skills = 1; any other skill combination = 0.	Calculated using O*NET 19.0
Low STEM/ High Soft	Dummy variable where occupations requiring Low STEM skills but High Soft skills = 1; any other skill combination = 0.	Calculated using O*NET 19.0
Control variables		
Education	Dummy variable where occupations requiring BA or higher are coded as 1; occupations requiring less than BA = 0.	Calculated using O*NET 19.0
Experience	Dummy variable where occupations requiring more than 1 year of experience are coded as 1; occupations requiring a year or less experience = 0.	Calculated using O*NET 19.0
OJT	Dummy variable where occupations requiring more than 3 months of on-the-job training are coded as 1; occupations requiring 3 months or less training = 0.	Calculated using O*NET 19.0
Employment	Natural log of occupational employment, transformed to adhere to regression assumptions about normalized distribution.	OES, May 2014

Table 4.2 Occupational Skill Categories and Education, Experience, and Training Requirements

	No. of occupations	Share (%)
Skill category		
High STEM/High Soft	182	23.8
High STEM/Low Soft	168	22.0
Low STEM/High Soft	155	20.3
Low STEM/Low Soft	259	33.9
Education		
BA or above	265	34.7
Less than BA	499	65.3
Experience		
1 year or less	325	42.5
More than 1 year	439	57.5
On-the-job training		
3 months or less	378	49.5
More than 3 months	386	50.5

NOTE: $N = 764$.
SOURCE: O*NET and OES (2014); author calculations.

Table 4.3 provides data on the number of people employed in each of the four KSA categories. Total U.S. employment was 131.8 million in 2014, according to the OES data. Of that number, 47.4 million workers (35.9 percent) were in occupations requiring High Soft skills, while only 36.7 million workers (27.8 percent) had jobs requiring High STEM KSAs. Although 27 more occupations were categorized as requiring High STEM/High Soft skills than Low STEM/High Soft skills, the latter category employed 21.6 percent more workers than the former. Occupations requiring High STEM and Low Soft KSAs accounted for the smallest share of employment by far, employing only 11.6 percent of the total U.S. workforce.

It's important to note that the employment captured as requiring High STEM skills represents a larger number of STEM occupations compared to the 100 occupations the U.S. Bureau of Labor Statistics identifies as being STEM occupations. As discussed previously, for this analysis, occupations classified as High STEM include those that require above-average technical and mechanical KSAs, as well as those requiring above-average knowledge of social science domains.

Table 4.3 U.S. Employment by STEM/Soft Category

	Low STEM	High STEM	Total
High Soft	25,990,470	21,366,660	47,357,130
	19.7%	16.2%	35.9%
	(N = 155)	(N =182)	(N = 337)
Low Soft	69,157,630	15,298,390	84,456,020
	52.5%	11.6%	64.1%
	(N = 259)	(N = 168)	(N = 427)
Total	95,148,100	36,665,050	131,813,150
	72.2%	27.8%	100%
	(N = 414)	(N = 350)	(N = 764)

SOURCE: O*NET and OES (2014).

Inclusion of occupations in the social sciences is consistent with the NSF definition of STEM, whereas the BLS does not include such occupations. The BLS estimated employment in STEM jobs to be 7.9 million in 2012, or roughly 6 percent of total U.S. employment, and it projected employment in STEM occupations to grow to 9 million by 2022 (Vilorio 2014).

Despite the considerable policy and media focus on High STEM jobs and STEM degrees, 95.1 million workers nationwide were employed in occupations requiring Low STEM skills. There are 2.6 Low STEM jobs for every High STEM job in the United States. Particularly concerning, 69.2 million workers (52.5 percent of all workers) were employed in jobs requiring both Low STEM and Low Soft skills.

Do STEM occupations pay more? The descriptive data reveal that occupations requiring a high intensity of Soft skills value such skills. The 350 occupations requiring above-average STEM KSAs paid a median wage of $53,775. The 337 occupations requiring above-average Soft KSAs paid a median wage of $64,570, more than $10,000 higher than the median for STEM-intensive occupations. The wage for the High Soft skilled occupations was similar to, but slightly less than, the $67,790 median wage for the 265 occupations requiring a bachelor's degree or higher. Figure 4.2 displays median wages for the four STEM/Soft categories, without controlling for differences in education, experience, and training requirements. Occupations requiring above-average STEM and above-average Soft skills paid the highest median wages, higher even than the median wage for occupations requiring at least a

Figure 4.2 Occupational Median Wage by STEM/Soft Category

	Low STEM	High STEM
High Soft	$57,360 ($N = 155$)	$72,220 ($N = 182$)
Low Soft	$29,500 ($N = 259$)	$41,300 ($N = 168$)

SOURCE: O*NET and OES (2014); author calculations.

four-year college degree. Moreover, Low STEM/High Soft occupations paid 38.9 percent more than occupations requiring High STEM/Low Soft skills ($57,360 vs. $41,300).

Table 4.4 provides data on the number and share of occupations in the four STEM/Soft categories that require at least a bachelor's degree. (The total number of occupations and employment for each category is provided in Table 4.3.) The results support the view that higher education is a proxy for higher skill; nearly three quarters of occupations with the highest skill requirements also required at least a bachelor's degree. A similar share of occupations requiring below-average STEM but above-average Soft skills also required a four-year college degree or more, but less than 7 percent of occupations in the High STEM/Low Soft category required a bachelor's degree or higher. This finding suggests a closer relationship between higher education and above-average Soft skills than above-average STEM skills. It may indicate that, for many employers, a bachelor's degree helps to signal the presence of hard-to-assess Soft skills.

Table 4.4 Share of Occupations by Skill Category Requiring Bachelor's Degree or Higher

Skill category	Number of occupations requiring BA+	Share of occupa-tions requiring BA+ (%)	Share of employment in occupations requiring BA+ (%)
High STEM/High Soft	132	72.5	49.5
High STEM/Low Soft	11	6.6	9.6
Low STEM/High Soft	112	72.3	60.6
Low STEM/Low Soft	8	3.1	3.0

SOURCE: O*NET and OES (2014).

The difference in the share of job holders with at least a bachelor's degree between the High STEM/High Soft and Low STEM/High Soft categories is also interesting. Only about half of High STEM/High Soft employment was in jobs requiring that level of education, as compared to 61 percent of Low STEM/High Soft employment in high-education occupations. The 132 High STEM/High Soft occupations requiring at least a bachelor's degree employ 10.6 million workers, whereas the 112 Low STEM/High Soft occupations requiring at least a bachelor's degree employ 15.8 million workers. Clearly, occupations requiring a higher level of education related to STEM employ far fewer workers than those requiring a higher level of education related to Soft skills. This may indicate differences in the nature of work, wherein technology-intensive activities demand fewer workers than people-intensive ones.

A linear regression analysis was conducted to test the explanatory power of the KSA variables in predicting median wage, controlling for variables related to education and experience requirements, as well as total occupational employment. The dummy variable meant to capture on-the-job training requirements was removed from the model, owing to the extremely high degree of correlation with the experience variable. Given that exploring an alternative measure of human capital that is both finer grained and broader based than the common proxy of educational attainment is one goal of this research, a two-stage hierarchical model was used to examine whether the STEM/Soft variables added explanatory power beyond what could be provided by using a dummy variable to indicate whether an occupation required a bachelor's degree or higher.

As can be seen in Table 4.5, the three STEM/Soft dummy variables were statistically significant, even after controlling for education, experience, and occupational employment. The model adding the KSA variables to the education, experience, and occupational employment variables showed improved explanatory power over the control variables alone, increasing R^2 by 0.12. Both models were significant at the $p < 0.001$ level. All of the variables indicating occupational requirements were positively significant at the $p < 0.001$ level; the employment variable was negatively significant.

Model 2 demonstrates how much better paid occupations are requiring some higher level of skill relative to those requiring both below-average STEM and below-average Soft skills. After controlling for differences in educational and experience requirements and occupational employment levels, occupations requiring High STEM and High Soft skills paid nearly $86,500 more than that of occupations requiring below-average STEM and below-average Soft skills.[1] Occupations requiring High STEM but Low Soft skills paid $44,920 more, and occupations requiring below-average STEM but above-average Soft skills paid $59,260 more. These results suggest that occupation-based human capital, measured as above- and below-average STEM and Soft KSAs, is a useful measure in predicting median wage. As hypothesized, the highest wages were in occupations requiring both above-average STEM and above-average Soft skills.

The regression analysis largely confirms the hypothesized relationships between occupational skill requirements and median wages. As theory predicts and casual observation suggests, occupations with the highest skill requirements—those requiring both above-average STEM and above-average Soft KSAs—pay the highest wages among the four categories, and occupations with the lowest skill demands—those requiring below-average STEM and below-average Soft KSAs—pay the lowest. The fact that the two remaining skill categories indicating at least some higher skill requirement pay a higher wage than the lowest category of skill confirms the assumption that higher skills tend to be rewarded with higher pay. However, given the current attention paid to STEM skills, it is interesting that the category indicating above-average Soft but below-average STEM skill requirements paid a higher median wage than occupations in the High STEM/Low Soft category.

Table 4.5 Regression Analysis Models of Relationship between Occupational Skill Sets and Log-Transformed Median Wage

Variables	Model 1		Model 2	
	Coefficient	t	Coefficient	t
Intercept	−0.75	−18.67***	−0.99	−24.09***
Occupations w/BA+	1.03	17.49***	0.55	8.09***
Experience	0.68	12.04***	0.40	7.36***
Ln employment	−0.07	−2.64**	−0.06	−2.40*
High STEM/High Soft	—	—	1.19	14.86***
High STEM/Low Soft	—	—	0.54	8.11***
Low STEM/High Soft	—	—	0.81	9.94***
	$R^2 = 0.5$		$R^2 = 0.62$	
	Adj. $R^2 = 0.5$		Adj. $R^2 = 0.61$	
	F (df) = 253.01 (3,760)***		F (df) = 201.2 (6,757)***	
			R^2 change = 0.12	
			F-change = 75.24***	

NOTE: $N = 764$; $*p \leq 0.05$; $**p \leq 0.01$; $***p \leq 0.001$.

What explains this somewhat unanticipated finding? One obvious difference between the two categories is that far more Low STEM/High Soft occupations require a higher level of educational attainment. As seen in Table 4.4, 72 percent of Low STEM/High Soft occupations (and 61 percent of employment) required a bachelor's degree or higher, compared to just 7 percent of High STEM/Low Soft occupations (and 10 percent of employment). In addition to educational differences, the higher median wage may be partly due to job hierarchy. Management occupations make up nearly 12 percent of the Low STEM/High Soft occupations; a higher position on the occupational hierarchy within companies tends to come with a higher wage. However, the higher wage for Low STEM/High Soft occupations highlights that skills other than STEM, such as communication, leadership, teamwork, decision making, and judgment, are valued in the workplace.

A sensitivity analysis that altered characteristics of the model revealed similar predictive ability whether the education dummy variable indicated associate's degree or higher (Adj. $R^2 = 0.62$) or master's degree or higher (Adj. $R^2 = 0.61$). However, the employment variable was no longer significant in the model using the master's and above dummy variable. Given that the occupational employment variable was included in the model to control for potential wage effects due to differences in the absolute employment size of occupations, an alternate model explored the predictive ability only among the 100 occupations employing the most workers. Using a dummy variable to limit the dataset to only those occupations employing 300,000 workers or more substantially improved the model's predictive ability (Adj. $R^2 = 0.78$). All the independent STEM/Soft variables, as well as the education and experience control variables, were positively significant.

The findings suggest that many individuals may be better served by efforts to improve their Soft skills instead of getting too exclusively caught up in the current focus on STEM. Occupations requiring above-average Soft skills accounted for a substantially larger share of employment (35.9 percent) than did High STEM occupations (27.8 percent). In addition, Low STEM/High Soft occupations paid considerably more than High STEM/Low Soft occupations. The findings also indicate the outsized impact of low-skill occupations. Low STEM/Low Soft occupations accounted for only about a third of all occupations but 52.5 percent of all U.S. employment. In other words, more than half of all workers held jobs that, by and large, paid considerably lower wages. The high-skill jobs that attract so much policy attention do command much higher wages, but they also demand far fewer workers.

Note

1. High STEM/High Soft occupations paid $e^{(1.19)}$ times the SD ($26,310) of the overall mean of the occupational median wage, or $86,483, more than the reference Low STEM/Low Soft group. The calculations for the other two categories were similar. High STEM/Low Soft paid $e^{(0.54)}$ times the SD, or $44,923, and Low STEM/High Soft paid $e^{(0.81)}$ times the SD, or $59,261.

5

The Intersection of Skill Demand and Regional Well-Being

A spate of initiatives at the local and state level has been directed at upgrading STEM skills in the workforce. Curricula in secondary and primary schools throughout the country have been realigned so that thousands of students are educated in learning environments that emphasize STEM knowledge over other subject matter, such as language, history, and art. Public resources in the form of grants and scholarships have been allocated to encourage more college students to pursue STEM fields. Universities, corporations, and foundations have developed programs to expose students to STEM career opportunities and provide STEM-specific training certification for teachers. Community colleges have expanded "stackable" short-term credentials and training partnerships with local employers. Local school districts have either turned their career and technical schools, formerly known as vocational programs, into a track for "problem" students or have curtailed programs in manufacturing and the trades as enrollments have dropped or funds were reallocated to support STEM programs (National Center for Education Statistics, n.d.; National Education Association 2012). Area workforce development agencies have supported STEM-aligned summer camp programs, apprenticeships, and leadership councils. State and local economic development organizations have created STEM strategic priorities.

Certainly, such initiatives are rooted in studies linking human capital investments to individual and societal gains. They have been guided by analyses and articles by university research centers and federal agencies (see Carnevale, Smith, and Melton 2011; National Science Board 2015) citing wage and employment advantages of STEM majors and STEM occupations. Reports and campaigns by various business and industry advocacy groups, such as the Business Roundtable and the Manufacturing Institute, calling for action to address STEM skill needs have shaped these initiatives. State and local political leaders have internalized largely national explorations of STEM gains and gaps to advance their own STEM education agendas as a path to growth and prosperity.

Yet, scant academic literature specifically assesses the impact of STEM on measures of economic growth, particularly at a subnational level. Rothwell (2013) offers a notable exception in revealing the contributions that "hidden" STEM occupations—meaning those requiring less than a college degree—make to regional economies.

Instead, the proliferation of state and local STEM initiatives reflects a set of assumptions: 1) Human capital investments in STEM that lead to higher worker wages will also generate higher economic growth. 2) Relatively similar investments in STEM development will yield expected returns regardless of differences in human capital deployment capacity. 3) Economic benefit from STEM investments seen at the national level will accompany similar investments at the subnational level.

Linking the methodology described in Chapters 3 and 4 to geographic employment data enables an exploration of these assumptions. As with Rothwell (2013), this analysis focuses on variation in occupational human capital concentrations found in regions, specifically metropolitan statistical areas (MSAs). MSAs were selected as the unit of analysis because they are the best available approximation of a labor market area. MSAs are geographic areas within which people both live and work, and they are defined by commuting patterns. Because they follow commuting patterns, they are not constrained by political borders, such as state and county lines. Exploring human capital at the subnational level requires recognition of this flow of workers. For purposes of study, MSAs offer the analytical advantage of capturing this frequent shifting across political jurisdictions that characterizes work life and home life. The mismatch between economic geography and political geography does make it difficult to coordinate public-sector investments that can benefit the economy and potential workers. The fact that MSAs often encompass multiple cities, municipalities, and at times states makes developing, coordinating, and implementing policies across jurisdictional boundaries difficult.

Despite these challenges, analysis of regional differences in human capital deployment and associated economic well-being offers critical insights for policymakers. The research in this chapter demonstrates that variation in regional human capital deployment—measured in terms of occupation-based skill requirements—better explains differences in regional economic well-being than do the more common education-based measures that reflect regional human capital development. This study specifically explores regional human capital deployment within the context of STEM KSAs and Soft KSAs. Findings call into question broad-stroke assumptions about more highly skilled and highly educated areas being places of better economic well-being, especially when measures other than regional median wage are used. As the analysis makes clear, higher concentrations of human capital are associated with better economic performance, but the relationship—at the regional level, at least—is far more complex and nuanced than current, simplified operationalizations of human capital theory suggest and regional policies that are largely imitative of national or other regional initiatives assume.

PUTTING REGIONAL STEM CONCENTRATIONS TO THE TEST

The reshaping of educational systems to emphasize STEM, the rise in policies designed to increase the number of workers with such skills, and evidence out of the business literature indicating employer demand for critical thinking, communication, and other "soft" skills suggest a series of four general hypotheses that an occupation-based measure of regional human capital can be used to test.

H1. Regions with a greater share of employment in occupations requiring High STEM and High Soft KSAs experience greater regional economic well-being than those that do not share in this attribute.

H2. Regions with a larger share of employment in occupations requiring High STEM but Low Soft skills experience greater economic well-being, but less than regions with larger concentrations of the highest skill occupations.

H3. Regions with greater shares of employment in occupations requiring Low STEM but High Soft KSAs experience economic benefit, but it is less than regions with a greater share of employment in High STEM occupations.

H4. Regions with a larger share of employment in occupations requiring both Low STEM and Low Soft skills experience weaker economic well-being than those with stronger human capital attributes.

Testing these hypotheses requires a method for deriving measures of occupation-based regional human capital and linking these measures to indicators of regional economic well-being. It was a relatively straight-forward process to apply the occupational skill categories described in Chapter 3 to occupational employment data available at the MSA level from OES. As noted earlier, O*NET is a national database, and its sample size is not large enough to assess possible regional variations in skill requirements for the same occupation. In other words, O*NET does not offer insight into whether there are skills expected of a machinist in Cleveland that are not required of a machinist in Birmingham. However, it is possible to explore variation in regional human capital based on differences in employment concentrations of skill sets (for a theoretical discussion of regionally different occupational mixes, see Markusen et al. 2008).

The use of Standard Occupational Classification (SOC) codes by both O*NET and OES enabled linking the occupation-based human capital categories described in Chapters 3 and 4 to MSA employment and wage data. Each region's share of employment in the four STEM/ Soft categories was calculated. Shared coding systems delineating MSAs and New England County and Town Areas (NECTAs) also facilitated matching OES data to demographic and socioeconomic data from the American Community Survey (ACS 2013).

Previous studies exploring O*NET data have tended to focus on the effects of regional human capital variation on wages or employment (Florida et al. 2012; Koo 2005; Maxwell 2008; Rothwell 2013; Scott 2009; Yakusheva 2010). However, a region's economic well-being goes beyond wages and employment. Change in gross regional product (GRP), total factor productivity, per capita income, poverty rate, and income inequality are all measures of economic well-being that are found in the economics and economic development literature (see, for example, Baum and Ma 2007; Benhabib and Spiegel 1994;

Chrisinger, Fowler, and Kleit 2012; Gottlieb and Fogarty 2003; Holzer 2008; Lerman 2008; Moretti 2004; Wolfe and Gertler 2004). Although the human capital literature largely suggests an across-the-board positive benefit to greater levels of education and skill, Andreason (2015) presented a more nuanced view, where increases in human capital, measured as the share of population with a bachelor's degree or higher, improved some regional economic indicators but had no effect on or worsened others. Given such mixed results across important measures of regional economic well-being, this research analyzed the effects of regional human capital variation on five separate measures: metropolitan median wage in 2014, percent change in GRP from 2009 to 2013 (in 2013 $), total factor productivity in 2013, per capita income in 2013, and poverty rate in 2013.

Figure 5.1 summarizes the hypothesized relationships between regional human capital deployment and regional economic well-being. Although Andreason's findings informed this research, the hypotheses reflect the assumed broad-based benefits of higher human capital concentrations; that is, regions with a larger share of employment in occupations requiring High STEM and High Soft KSAs will pay the highest wages, see the greatest GRP growth, have the highest productivity, enjoy the highest per capita incomes, and experience the lowest rates of poverty. Given the emphasis on STEM skills broadly, regions with a larger share of employment in occupations requiring High STEM, despite Low Soft skill requirements, will pay higher wages, see greater GRP growth, have higher productivity, enjoy higher per capita incomes, and experience lower poverty rates than regions with lower shares of such employment, but less than regions with larger concentrations of the highest skill occupations. Regions with a larger share of employment in occupations requiring Low STEM but High Soft skills will pay higher wages, see greater GRP growth, have higher productivity, enjoy higher per capita incomes, and experience lower poverty rates than regions with lower shares of such employment, but less than for regions with larger concentrations of employment in High STEM occupations. Regions with a larger share of employment in occupations in the lowest skill category, Low STEM/Low Soft, will pay the lowest wages, see the lowest GRP growth, have the lowest productivity, have the lowest per capita incomes, and experience the highest poverty rates.

Figure 5.1 Hypothesized Effects of Occupation-Based Human Capital on Five Measures of Regional Economic Well-Being

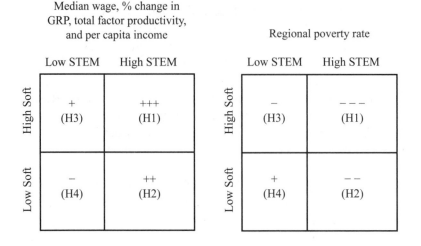

NOTE: The plus signs in the left panel represent the relative strength of the hypothesized positive relationship between higher occupational skill requirements and the four listed measures of regional economic well-being (the minus sign represents a weakness). The minus signs in the right panel represent the relative strength of the hypothesized inverse relationship, where higher occupational skill requirements are theorized to be associated with lower regional poverty rates (the plus represents a relatively higher poverty rate).

Many factors, such as population size and labor force participation rates, have been shown to affect regional economic well-being. The academic literature indicates that many of these same factors are also associated with human capital accumulation. For example, larger cities tend to attract people with higher skills (e.g., Combes et al. 2012; Elvery 2010; Glaeser and Saiz 2003; Gould 2007; Moretti 2004). In general, areas with better-educated residents tend to experience better economic performance (e.g., Baum and Ma 2007; Ehrlich 2007; Feser 2003; Glaeser and Saiz 2003; Goldin and Katz 2010; Gottlieb and Fogarty 2003; Lucas 1988, 2009; Markusen et al. 2008; Moretti 2004; Nelson and Phelps 1966; Wolfe and Gertler 2004). Areas with better-educated residents also tend to attract new residents (Bartik 1993; Black and Henderson 1999; Glaeser and Mare 2001; Partridge and Rickman 2003; Simon 1998; Simon and Nardinelli 2002). These insights shaped the development of a model to test the value of an occupation-based measure of regional human capital.

In addition, the literature suggests several control variables. Median house value helps to control for regions experiencing higher wages, higher growth, and often higher costs of living (Capozza et al. 2002; Glaeser and Saiz 2003). Areas where a larger share of working-age adults are actually working should see greater economic performance than those regions where higher shares of eligible workers are idle (Glaeser and Saiz 2003; Kodrzycki and Muñoz 2013). The share of employment engaged in manufacturing helps to control for the effects of industry mix on economic performance (Blumenthal, Wolman, and Hill 2009; Glaeser and Saiz 2003; Kodrzycki and Muñoz 2013).

Table 5.1 lists the variables incorporated in a series of regression analyses, their definitions, and sources. It should be noted that the effects of the Great Recession, which began in December 2007 and ended in June 2009, were still impacting the dependent and control variables used in this analysis. The GRP comparison begins the year the Great Recession ended. To be able to access ACS data at the regional level, the five-year estimates were used. This means that the 2013 estimate includes 60 months of data collection from January 1, 2009, to December 31, 2015. Moreover, the OES surveys 200,000 establishments every six months over a three-year cycle. As such, the May 2014 release includes wage and employment data from November 2011.

HUMAN CAPITAL MEASURES AND ECONOMIC PERFORMANCE VARY WIDELY ACROSS REGIONS

Table 5.2 provides the mean, standard deviation, coefficient of variation, minimum, and maximum for the variables. The mean median wage across all regions is displayed in the table even though the wage variable was log transformed for the regression analyses to correct for the skew in its distribution across metropolitan areas. In addition, for the regression analyses, all variables were standardized by calculating their z-scores for ease of interpretation due to different units of measurement. However, the descriptive statistics presented in Table 5.2 reflect each variable's measurement before transformation for ease of discussion.

Table 5.1 Variable Definitions and Data Sources for Regional Human Capital Analysis

Variable	Definition	Source
Dependent variables		
Median wage, 2014	Natural log of MSA median wage for all occupations	OES, May 2014
% Change in GRP, 2009–2013	Percent change in gross regional product, 2009–2013	Calculated using Moody's Analytics; calculated in 2013 $
Productivity, 2013	MSA GRP divided by total MSA employment, 2013	Calculated using Moody's Analytics
Per capita income, 2013	MSA per capita income for the previous 12 months in 2013 dollars	ACS 5-year estimate, 2013
Poverty rate, 2013	Share of MSA population below the poverty line, 2013	ACS 5-year estimate, 2013
Independent variables		
High STEM/High Soft employment	Share of MSA employment in occupations requiring both above-average STEM and above-average Soft skills	Calculated using O*NET 19.0 and OES, May 2014
High STEM/Low Soft employment	Share of MSA employment in occupations requiring above-average STEM but below-average Soft skills	Calculated using O*NET 19.0 and OES, May 2014

Low STEM/High SOFT employment	Share of MSA employment in occupations requiring below-average STEM but above-average Soft skills	Calculated using O*NET 19.0 and OES, May 2014
Low STEM/Low SOFT employment	Share of MSA employment in occupations requiring both below-average STEM and below-average Soft skills	Calculated using O*NET 19.0 and OES, May 2014
Control variables		
Population change (%), 2010–2013	Percent change in MSA population, 2010–2013	Calculated using ACS 5-year estimate, 2013, and 2010 Census
Labor force participation rate, 2013	Share of the MSA population age 16 and over in the labor force, 2013	Calculated using ACS 5-year estimate, 2013
Manufacturing employment (%), 2013	Share of the MSA total employment in manufacturing, 2013	Calculated using ACS 5-year estimate, 2013
Ratio of region to U.S. median house value, 2013	Ratio of MSA owner-occupied median house value to U.S. median of $160,000	Calculated using ACS 5-year estimate, 2013
% Population with BA or higher, 2013	Share of the MSA population age 25 and over with a BA degree or higher, 2013	Calculated using ACS 5-year estimate, 2013

Table 5.2 Descriptive Statistics of the Variables[a]

Variable	Mean	Std. dev.	CV	Min.	Max.
% High STEM/High Soft employment	13.1	3.1	0.23	5.0	25.4
% High STEM/Low Soft employment	9.5	2.6	0.27	4.1	25.1
% Low STEM/High Soft employment	15.9	3.1	0.20	8.1	26.2
% Low STEM/Low Soft employment	48.4	4.3	0.09	34.9	62.3
% Population change, 2010–2013	2.2	2.4	1.09	–4.6	11.0
% Labor force participation, 2013	63.7	4.9	0.08	44.1	75.3
% Employment in manufacturing, 2013	11.1	5.3	0.48	2.1	36.5
Region/U.S. median house value, 2013	1.1	0.6	0.50	0.0	4.6
% Population with BA or higher, 2013	26.9	8.3	0.31	11.9	58.3
Median wage ($), 2014	33,624	4,694	0.14	22,780	57,430
% Change in GRP, 2009–2013	6.5	8.9	1.37	–9.2	70.0
Total factor productivity ($), 2013	102,071	24,325	0.24	63,671	215,705
Per capita income ($), 2013	41,745	8,514	0.20	23,073	87,897
% Population below poverty line, 2013	15.8	4.3	0.27	5.5	34.8

NOTE: $N = 395$, except for per capita income, GRP, and productivity (384) and per capita income (394).

[a] Descriptive statistics are in raw data for ease of understanding; for the analysis, population change was measured as a dummy variable, median wage was calculated as its natural logarithm, and all variables were standardized by calculating z-scores.

The data show wide regional variation, both in terms of economic performance and in terms of human capital, whether measured as advanced education or by the high/low skill categories developed earlier:

- The gap between the metropolitan regions with the highest and the lowest median wages was nearly $35,000.

- Per capita incomes in the lowest performing metropolitan regions were little more than one quarter that of those in the highest performing regions.

- Although the metropolitan regions, on average, experienced tepid, but positive, five-year growth in GRP (measured in 2013 $), some regions saw their economies shrink while others surged.

- Total factor productivity (the ratio of GRP to employment) was a little more than $102,000 across all metropolitan regions in the sample, but the highest performing region had a total factor productivity that was more than 3.5 times that of the lowest performing MSA.

- Poverty in the lowest performing metropolitan region was more than double the average for all regions.

- Large regional differences in skill concentrations exist. The metropolitan region with the highest share of High STEM/High Soft employment had five times that of the region with the lowest share of such employment. The MSA with the highest share of High STEM/Low Soft employment had more than six times that of the region with the lowest share of such employment. The MSA with the highest share of Low STEM/High Soft employment had more than three times that of the region with the lowest share of such employment.

- Slightly less than half of employment across metropolitan regions, on average, was in Low STEM/Low Soft occupations, but some regions had as much as 60 percent of their workers in low-skill jobs.

- As wide as these occupational skill gaps were, they were not as great as the divide in educational attainment. Although, on average, slightly more than a quarter of each metropolitan region's population age 25 or over had a bachelor's degree or higher, the gap between the regions with the highest and lowest share was 46 percentage points.

Although dividing a region's share of employment into four quadrants indicating occupational skill requirements could be expected to introduce collinearity into the model, the four categories do not total 100 percent of regional employment. This is due to the fact that not all occupations have been mapped by O*NET; the OES survey does not include self-employed workers; certain occupations could not be matched or lacked sufficient data; and the OES suppresses data at the detailed occupational level if inclusion of the data may reveal specific establishments in an MSA. Although the four quadrants did capture more than 95 percent of regional employment for some MSAs, they captured little more than two-thirds in others. The average share of regional employment accounted for by the four skill categories was 86.9 percent.

TESTING THE INFLUENCE OF THE STEM/SOFT OCCUPATION-BASED MEASURES ON REGIONAL ECONOMIC WELL-BEING

Whether the wide variation in regional human capital, measured as occupation-based skill sets, helps to explain the wide variation in observed regional economic well-being was tested through a series of five regression analyses. As noted previously, human capital theory posits that areas with greater levels of human capital—whether defined as educational attainment or occupational skill—will see greater economic benefit than regions with lower levels of human capital.

One goal of this research was to explore occupational skill sets matched to political and mainstream rhetoric as a measure of regional human capital. Another goal was to test whether such a measure would have greater explanatory power than the commonly used human capital measure, share of a region's population with a bachelor's degree or higher. As such, a multistage approach, allowing each set of variables of interest to be entered separately, was adopted to explore whether the occupation-based skill set variables improved explanatory power. For each dependent variable, Model 1 shows results for four control variables: labor force participation rate in 2013; the share of regional employment in manufacturing in 2013; house value ratio, which is defined as the median value of a house in an MSA divided by the median house value in the United States in 2013; and the population growth rate from 2010 to 2013, measured as the percent change in population. Model 2 adds the educational variable of interest—the proportion of the regional population age 25 and older with at least a bachelor's degree in 2013—to the control variables. Model 3 substitutes the share of employment in occupations in each of the four STEM/Soft categories for the education variable. (As discussed above, the STEM/Soft skill categories do not account for all employment in an MSA, the sum varies across MSAs, and inclusion of all four occupational skill variables does not result in undue levels of multicollinearity.)

The supply of workers with a level of educational attainment would be assumed to reflect occupational demand for workers with a stock of knowledge, skills, and abilities. Indeed, this assumption is behind the common use of education as a proxy for human capital. Correlation between the regional education variable and the occupational skill variables precluded using the two different methods of measuring human capital in the same analysis. However, an alternative approach to exploring the relationship between educational attainment levels and occupational requirements is discussed in Chapter 6.

Model of Occupational Skill Measures Explains 80 Percent of Wage Variation

Table 5.3 presents the results from the three linear regression models of the determinants of the natural logarithm of metropolitan area median wages in 2014. The primary focus of this research is on the four STEM/Soft variables, which are the share of regional employment in each category examined in Model 3. In brief, Model 1 demonstrates that three of the four control variables included in this stage were significant and explained more than half of the variation in the natural logarithm of regional median wage (Adj. $R^2 = 0.58$). Population change was the only variable shown to have a negative effect on the median regional wage. Model 2 added the share of population attaining a bachelor's degree or higher to the model. This variable was significant, but it only modestly improved the model's explanatory power (Adj. $R^2 = 0.62$). As theory predicts, the education variable was positively associated ($b = 0.29$) with regional median wage, as were two of the four control variables.

Substituting the four STEM/Soft variables for the educational attainment variable in Model 3 significantly improved the explanatory power of the model (Adj. $R^2 = 0.80$), and all four of the STEM/Soft KSA variables had a statistically significant ($p < 0.001$) relationship with median wage. The share of regional employment in High STEM/High Soft occupations was positively associated with median wage ($b = 0.27$), as were the share of High STEM/Low Soft employment ($b = 0.11$) and the share of Low STEM/High Soft employment ($b = 0.15$). The share of regional employment in Low STEM/Low Soft occupations was negatively related to regional median wage ($b = -0.28$). The share of MSA employment in Low STEM/Low Soft occupations had an effect on the regional median wage that was similar in size but opposite in effect to the share of employment in High STEM/High Soft occupations. The effects of the share of regional employment in High STEM/Low Soft occupations and Low STEM/High Soft occupations were substantially

Table 5.3 Regression Models of the Relationship between Occupational Skill Sets and the Natural Logarithm of Regional Median Wage in 2014

Variables	Model 1		Model 2		Model 3	
	Coefficient	t	Coefficient	t	Coefficient	t
Intercept	0.00	0.00	0.00	0.00	0.00	0.00
Labor force participation rate	0.40	10.43***	0.29	6.89***	0.20	6.99***
Manufacturing employment (%)	−0.02	−0.40	0.03	0.71	0.05	1.78
Region/U.S. median house value	0.57	15.61***	0.45	11.18***	0.50	18.34***
Population change, 2010–2013	−0.17	−4.44***	−0.20	−5.25***	−0.18	−6.36***
Share of pop. BA or higher	—	—	0.29	6.13***	—	—
High STEM/High Soft employment	—	—	—	—	0.27	7.87***
High STEM/Low Soft employment	—	—	—	—	0.11	4.53***
Low STEM/High Soft employment	—	—	—	—	0.15	4.14***
Low STEM/Low Soft employment	—	—	—	—	−0.28	−11.16***
	$R^2 = 0.58$		$R^2 = 0.62$		$R^2 = 0.80$	
	Adj. $R^2 = 0.58$		Adj. $R^2 = 0.62$		Adj. $R^2 = 0.80$	
	F (df) = 136.32 (4, 390)***		F (df) = 126.8 (5, 389)***		F (df) = 192.7 (8, 386)***	
			R^2 change = 0.04		R^2 change = 0.22	
			F-change = 37.58***		F-change = 104.45***	

NOTE: $N = 395$ MSAs and NECTAs; *$p \leq 0.05$; **$p \leq 0.01$; ***$p \leq 0.001$.

smaller. Three of the control variables were also statistically significant ($p < 0.001$), with population change being the only one with a negative association.

Higher Employment in High STEM/Low Soft Occupations Linked to GRP Growth

Table 5.4 presents the results from the three linear regression models designed to explain the percent change in gross regional product (GRP) from 2009 to 2013 using the multistage approach employed in the previous section. Model 1 demonstrates that the share of manufacturing employment in an MSA and population growth rate from 2010 to 2013 were both positively associated with change in GRP. The high t-statistic for the manufacturing variable relates to the Great Recession. The automobile industry and its substantial supply chain collapsed between 2008 and 2009 and had been revived by 2013. The model was significant but explained less than a quarter of the regional variation in change in GRP (Adj. $R^2 = 0.21$). Adding the share of population with a bachelor's degree or higher (Model 2) only slightly improved the explanatory power (Adj. $R^2 = 0.23$). Somewhat surprisingly, the education variable was negatively associated ($b = -0.19$) with change in GRP.

Model 3, which substitutes the four occupational variables for the education variable, substantially improved the explanatory power of the equation (Adj. $R^2 = 0.39$) but still explained far less than half of regional variation in change in GRP from 2009 to 2013. Three of the STEM/Soft KSA variables had a statistically significant relationship with change in GRP, but only the share of regional employment in High STEM/Low Soft occupations was positively associated with it ($b = 0.37$). The other two significant human capital variables had a negative relationship with GRP growth. As human capital theory suggests, a higher share of employment in Low STEM/Low Soft occupations ($b = -0.21$) was a drag on regional GRP growth, but so was a higher share of regional employment in High STEM/High Soft occupations ($b = -0.17$). This unexpected finding would seem to undercut the theorized straightforward relationship between higher concentrations of human capital and regional economic growth.

Table 5.4 Regression Models of the Relationship between Occupational Skill Sets and Percent Change in Gross Regional Product from 2009 to 2013

Variables	Model 1		Model 2		Model 3	
	Coefficient	t	Coefficient	t	Coefficient	t
Intercept	0.00	−0.01	0.00	−0.04	0.00	0.02
Labor force participation	0.09	1.61	0.17	2.75**	0.06	1.22
Manufacturing employment (%)	0.29	5.59***	0.26	5.03***	0.20	4.03***
Region/U.S. median house value	−0.07	−1.45	0.01	0.11	0.06	1.27
Population change, 2010–2013	0.45	8.27***	0.46	8.58***	0.39	7.89***
Share of pop. BA or higher	—	—	−0.19	−2.82**	—	—
High STEM/High Soft employment	—	—	—	—	−0.17	−2.89**
High STEM/Low Soft employment	—	—	—	—	0.37	8.53***
Low STEM/High Soft employment	—	—	—	—	−0.01	−0.19
Low STEM/Low Soft employment	—	—	—	—	−0.21	−4.76***
	$R^2 = 0.22$		$R^2 = 0.24$		$R^2 = 0.40$	
	Adj. $R^2 = 0.21$		Adj. $R^2 = 0.23$		Adj. $R^2 = 0.39$	
	F (df) = 26.77 (4, 379)***		F (df) = 18.15 (5, 378)***		F (df) = 31.12 (8, 375)***	
			R^2 change = 0.02		R^2 change = 0.18	
			F-change = 7.97**		F-change = 27.88***	

NOTE: N = 384 MSAs and NECTAs; *$p \leq 0.05$; **$p \leq 0.01$; ***$p \leq 0.001$.

Three of Four Occupation Human Capital Variables Affect Regional Variation in Productivity Positively

Table 5.5 presents the results from the three linear regression models that estimate 2013 total factor productivity based on the multistage regression design employed above. Model 1 demonstrates that three of the four control variables—metropolitan labor force participation rates in 2013, relative metropolitan area house value in 2013, and the percent change in population from 2009 to 2013—were positive and significantly associated with the dependent variable. This equation explained a little more than a third (Adj. $R^2 = 0.36$) of the variation in regional total factor productivity in 2013. Model 2 demonstrates that adding the metropolitan area's share of the adult population who attained at least a bachelor's degree did not improve the explanatory power of the equation (Adj. $R^2 = 0.35$).

The third model substitutes the four occupational requirement variables for the educational attainment variable, significantly improving the explanatory power of the equation (Adj. $R^2 = 0.60$). All four of the STEM/Soft KSA variables were statistically significant, with three of them positively associated with regional productivity. This repeats the pattern of association in Table 5.3, where the regional median wage was the dependent variable. The share of regional employment in Low STEM/Low Soft occupations was the only variable in the model with a negative influence on regional productivity ($b = -0.18$). Based on the coefficients, the share of regional employment in High STEM/Low Soft occupations ($b = 0.40$) had by far the greatest impact on regional productivity among the human capital measures.

High STEM/Low Soft and Low STEM/High Soft Skills Raise Per Capita Incomes

Table 5.6 presents the results from the three linear regression models that compose the multistage modeling approach on regional per capita income. Model 1 shows that three of the four control variables were significantly associated with the dependent variable. The labor force participation rate and regional relative median house value were positive, and the region's share of manufacturing employment was negative. The educational attainment variable, the share of population 25

Table 5.5 Regression Models of the Relationship between Occupational Skill Sets and Regional Total Factor Productivity in 2013

Variables	Model 1		Model 2		Model 3	
	Coefficient	t	Coefficient	t	Coefficient	t
Intercept	0.00	0.06	0.00	0.05	0.01	0.20
Labor force participation	0.20	4.18***	0.21	3.76***	0.02	0.49
Manufacturing employment (%)	0.04	0.92	0.04	0.88	0.04	0.95
Region/U.S. median house value	0.46	10.04***	0.46	8.82***	0.47	12.23***
Population change, 2010–2013	0.13	2.69**	0.13	2.69***	0.07	1.65
Share of pop. BA or higher	—	—	−0.01	−0.16	—	—
High STEM/High Soft employment	—	—	—	—	0.13	2.69**
High STEM/Low Soft employment	—	—	—	—	0.40	11.50***
Low STEM/High Soft employment	—	—	—	—	0.20	3.97***
Low STEM/Low Soft employment	—	—	—	—	−0.18	−4.94***
	$R^2 = 0.36$		$R^2 = 0.36$		$R^2 = 0.61$	
	Adj. $R^2 = 0.36$		Adj. $R^2 = 0.35$		Adj. $R^2 = 0.60$	
	F (df) = 53.83 (4, 379)***		F (df) = 42.96 (5, 378)***		F (df) = 72.42 (8, 375)***	
			R^2 change = 0.00		R^2 change = 0.25	
			F-change = 0.03		F-change = 58.41***	

NOTE: N = 384 MSAs and NECTAs; *$p \leq 0.05$; **$p \leq 0.01$; ***$p \leq 0.001$.

Table 5.6 Regression Models of the Relationship between Occupational Skill Sets and Regional Per Capita Income in 2013

Variables	Model 1		Model 2		Model 3	
	Coefficient	t	Coefficient	t	Coefficient	t
Intercept	0.01	0.29	0.01	0.31	0.01	0.30
Labor force participation	0.33	8.88***	0.26	6.24***	0.24	6.23***
Manufacturing employment (%)	−0.08	−2.25**	−0.06	−1.55	−0.07	−1.87
Region/U.S. median house value	0.56	15.84***	0.48	12.05***	0.56	15.61***
Population change, 2010–2013	−0.07	−1.73	−0.08	−2.18*	−0.09	−2.47***
Share of pop. BA or higher	—	—	0.19	3.97***	—	—
High STEM/High Soft employment	—	—	—	—	0.07	1.50
High STEM/Low Soft employment	—	—	—	—	0.18	5.48***
Low STEM/High Soft employment	—	—	—	—	0.12	2.67**
Low STEM/Low Soft employment	—	—	—	—	−0.08	−2.48*
	$R^2 = 0.59$		$R^2 = 0.60$		$R^2 = 0.64$	
	Adj. $R^2 = 0.58$		Adj. $R^2 = 0.60$		Adj. $R^2 = 0.64$	
	F (df) = 137.38 (4, 389)***		F (df) = 117.24 (5, 388)***		F (df) = 86.95 (8, 385)***	
			R^2 change = 0.02		R^2 change = 0.06	
			F-change = 15.78***		F-change = 15.72***	

NOTE: $N = 394$ MSAs and NECTAs; *$p \leq 0.05$; **$p \leq 0.01$; ***$p \leq 0.001$.

and above that earned at least a bachelor's degree, was added in Model 2. Including this measure of regional human capital improved explanatory power only slightly (Adj. R^2 = 0.60 vs. 0.58 in Model 1). The education variable was positively associated with regional per capita income (b = 0.19).

In the third version of the model, the four STEM/Soft skill variables were substituted for the educational attainment variable. This substitution resulted in a modest increase in the explanatory power of the regression equation (Adj. R^2 = 0.64). Three of the STEM/Soft KSA variables had a statistically significant relationship with per capita income. The share of regional employment in High STEM/Low Soft occupations (b = 0.18) and the share of employment in Low STEM/High Soft occupations (b = 0.12) were positively associated with per capita income, whereas the variable that measures the share of regional employment in Low STEM/Low Soft occupations (b = −0.08) was negatively associated with regional per capita income. Surprisingly, given all the attention paid to STEM jobs, the occupational human capital variable that captures most employment of scientists and engineers (High STEM/High Soft) did not have a statistically significant effect on regional per capita income.

High STEM/Low Soft Employment Linked to Lower Regional Rates of Poverty

Table 5.7 presents the results from the three linear regression equations that explain the determinants of the share of a region's population with incomes that are below the poverty line. Model 1 shows that two of the control variables were negatively associated with poverty rates; lower poverty rates were seen in regions with higher labor force participation rates (b = −0.49) and in regions where median house values are above the U.S. median (b = −0.36). Regions with higher rates of population growth from 2010 to 2013 tended to have higher rates of poverty (b = 0.15). However, this model explained less than half of regional variation in poverty rates (Adj. R^2 = 0.44). Model 2 demonstrates that adding the share of regional population that had attained at least a bachelor's degree as an independent variable did not improve explanatory power. The education variable had no statistically significant association with regional poverty rates. Substituting the four occupation-based

Table 5.7 Regression Models of the Relationship between Occupational Skill Sets and Regional Poverty Rates in 2013

Variables	Model 1		Model 2		Model 3	
	Coefficient	t	Coefficient	t	Coefficient	t
Intercept	0.00	0.00	0.00	0.00	0.00	0.00
Labor force participation	−0.49	−11.04***	−0.50	−9.91***	−0.46	−9.76***
Manufacturing employment (%)	−0.07	−1.6	−0.07	−1.53	−0.03	−0.69
Region/U.S. median house value	−0.36	−8.56***	−0.37	−7.55***	−0.40	−8.96***
Population change, 2010–2013	0.15	3.21***	0.14	3.16**	0.19	4.11***
Share of pop. BA or higher	—	—	0.01	0.23	—	—
High STEM/High Soft employment	—	—	—	—	−0.04	−0.67
High STEM/Low Soft employment	—	—	—	—	−0.18	−4.55***
Low STEM/High Soft employment	—	—	—	—	0.00	0.04
Low STEM/Low Soft employment	—	—	—	—	0.00	−0.05
	$R^2 = 0.44$		$R^2 = 0.44$		$R^2 = 0.47$	
	Adj. $R^2 = 0.44$		Adj. $R^2 = 0.44$		Adj. $R^2 = 0.46$	
	F (df) = 77.67 (4, 390)***		F (df) = 61.99 (5, 389)***		F (df) = 43.53 (8, 386)***	
			R^2 change = 0.00		R^2 change = 0.03	
			F-change = 0.05		F-change = 5.67***	

NOTE: $N = 395$ MSAs and NECTAs; *$p \leq 0.05$; **$p \leq 0.01$; ***$p \leq 0.001$.

human capital measures for the educational attainment variable in the third model improved explanatory power, but only slightly (Adj. R^2 = 0.46). Only one of the four skill variables was significant: the High STEM/Low Soft variable was negatively associated with poverty level ($b = -0.18$), meaning that regions with a larger share of such employment tended to have lower poverty rates.

WHAT THE FINDINGS MEAN

Understanding what the findings mean in real terms requires converting the standardized coefficients of the statistically significant occupation-based human capital measures back into their original units of measurement. Holding all other variables constant (Table 5.8):

- Regions that had a 1 standard deviation (3.1 percentage points) larger share of employment in High STEM/High Soft occupations had a regional median wage that was roughly $6,131 higher and total factor productivity that was $3,138 greater, but the change in GRP from 2009 to 2013 was 1.5 percentage points lower than the average MSA.

- Regions that had a 2.6 percentage point (1 standard deviation) larger share of employment in High STEM/Low Soft occupations had a higher median wage ($5,250 greater), GRP (3.3 percentage points greater), total factor productivity ($9,779 higher), and per capita income ($1,507 more), and a regional poverty rate that was 0.8 percentage point lower.

- Regions where the share of regional employment in Low STEM/High Soft occupations was 3.1 percentage points (1 standard deviation) larger had a higher regional median wage ($5,443 more), total factor productivity ($4,914 more), and per capita income ($1,056 higher).

- Regions where the share of regional employment in Low STEM/Low Soft occupations was 4.3 percentage points (1 standard deviation) higher had a lower regional median wage ($3,548 less), GRP growth (1.9 percentage points less), total factor productivity ($4,281 lower), and per capita income ($690 less).

Table 5.8 Summary of the Impact of a One Standard Deviation Increase in the Share of Employment in a Specific Occupational Group on Five Measures of Regional Economic Performance

	Occupational group			Five dependent or outcome variables				
Category	Mean share of regional employment (%)[a]	SD (% point)	Median wage ($)	Gross regional product (% pt.)	Total factor productivity ($)	Per capita income ($)	Poverty rate (%)
High STEM/High Soft	13.1	3.1	6,131	−1.5	3,138		
High STEM/Low Soft	9.5	2.6	5,250	3.3	9,779	1,507	−0.8
Low STEM/High Soft	15.9	3.1	5,443		4,914	1,056	
Low STEM/Low Soft	48.4	4.3	−3,548	−1.9	−4,281	−690	

NOTE: A blank cell indicates that the impact was not statistically significant from having no impact.

[a] The percentages do not add up to 100% because not all occupations have been mapped by O*NET; the OES survey does not include self-employed workers; certain government occupations are not included in this analysis; and the OES suppresses data at the detailed occupational level if inclusion of the data may reveal specific establishments in an MSA.

The results from the models demonstrate that some of the control variables had a large effect on regional economic well-being. To provide some context, holding all other variables in Model 3 equal (Table 5.9):

- A 1 standard deviation (4.9 percentage points) increase in labor force participation was associated with a $5,750 increase in regional median wage, a $2,001 increase in per capita income, and a 2.0 percentage point decrease in poverty.

- A 1 standard deviation (0.57) increase in the ratio of MSA median house value to U.S. median house value was associated with a regional median wage that was $7,747 higher, total factor productivity that was $11,433 higher, per capita income that was $4,742 higher, and a poverty rate 1.7 percentage points lower. It should be noted that it is difficult to identify the direction of the relationship because, for example, higher wages may lead to higher house values, and higher house values may mean workers require higher wages. Given that median house value is being used in this analysis as a proxy for regional cost of living, the poverty findings should be interpreted cautiously: a smaller share of residents in regions with comparatively higher median house values may fall below the national poverty level, but their above-poverty wages may simply reflect a higher cost of living and mask the relative impoverishment of workers.

- A 5.3 percentage point increase (1 standard deviation) in the share of regional employment engaged in manufacturing was associated with GRP growth that was 1.8 percentage points higher.

- Regions with a 2.4 percentage point increase in population from 2010 to 2013 (1 standard deviation) had a regional median wage that was $3,917 lower, GRP growth that was 3.5 percentage points higher, per capita incomes that were $775 lower, and poverty rates that were 0.8 percentage point higher.

SUMMARY

The findings indicate that exploring regional human capital through occupational skill requirements enhances understanding of the opera-

Table 5.9 Summary of the Impact of a One Standard Deviation Increase in the Specified Control Variable on Five Measures of Regional Economic Performance

Variable			Five dependent or outcome variables			
Control variable	One SD (% point)	Median wage ($)	Gross regional product (% pt.)	Total factor productivity ($)	Per capita income ($)	Poverty rate (%)
Labor force participation rate, 2013	4.9	5,750			2,001	−2.0
Relative home price: MSA to U.S.	0.57[a]	7,747		11,433	4,742	−1.7
Share of employment in manufacturing	5.3		1.8			
Population growth rate, 2010–2013	2.4	−3,917	3.5		−775	0.8

NOTE: A blank cell indicates that the impact was not statistically significant from having no impact.

[a] This quantity has no unit because it is a ratio.

tions of the labor market. The results support the importance of STEM skills to regional economic performance, but they demonstrate that the overriding focus on STEM degrees is too narrow. Regions with a higher share of High STEM/High Soft employment, the category encompassing scientists, engineers, and software applications developers, had higher wages and higher productivity, but regions that experienced more broad-based economic benefit for the study period were those with a higher share of High STEM/Low Soft employment. In other words, the broadest improvement in economic well-being was in regions with a higher share of workers such as computer programmers, electro-mechanical technicians, and CNC machine programmers. Yet, the results also indicate that STEM is not the only path to better regional economic well-being. Regions with a higher share of employment in occupations requiring High Soft skills, whether in combination with High STEM skills or not, had higher median wages and higher productivity.

Clearly, the demand-based human capital deployment variables reflecting occupational skill requirements offer an improved substitute over the rather blunt, supply-based proxy of educational attainment. However, it is reasonable to question whether four broad categories of occupational skill requirements are themselves too blunt a tool for adequately capturing the expansive concept of human capital. It is also reasonable to question the extent to which human capital concentrations and deployment, and the associated benefits, are shaped by the context of regions. These questions guide the analysis in the chapters that follow.

6

A Test of Size, Schooling, and Context

The regression analyses presented in Chapter 5 make clear that a broader conceptualization of human capital—measured in terms of occupational skill requirements—demonstrated greater and broader explanatory power than did its well-established proxy, which measured the share of the population with a bachelor's degree or higher. However, education and skill are related. This relationship was apparent in the moderately high correlation between the occupational and educational measures that precluded entering them together into the regression models presented in Chapter 5.

Population is also related to education and skill. This relationship was evident in the relatively high correlations between population and education and population and skill that prevented the use of MSA population level as a control variable in the regression models in Chapter 5.

This chapter presents methods for testing the sensitivity of the occupation-based human capital variables within the important contexts of education and population.

EDUCATIONAL ATTAINMENT AND OCCUPATIONAL SKILL: THE INTERPLAY OF HUMAN CAPITAL DEVELOPMENT AND DEPLOYMENT

Despite the multicollinearity challenge observed in the preceding chapter, the interplay between human capital development (educational attainment) and human capital deployment (the STEM-Soft occupational skill requirements) would seem to be germane to understanding variations in regional economic well-being. Andreason (2015) demonstrated the explanatory value and versatility of a different measure of human capital development—change in the share of population with a bachelor's degree. This measure was shown to offer an advantage in the context of this analysis over the educational attainment variable reflecting the share of a region's population with a bachelor's degree or higher in that the change in the share of highly educated residents did not have an unacceptably high level of multicollinearity with the occupational skill and control variables.

Similar to the results reported in Andreason (2015), the 395 MSAs examined in this analysis varied considerably in their ability to grow their share of the adult population with at least a bachelor's degree. Some regions expanded their college-educated population by more than 4 percent between 2009 and 2013, while others saw their college-educated population shrink by more than 1.5 percent. On average, the college-educated population across regions increased by 1 percent over the five-year span.

Although Andreason's work found mixed economic results for metropolitan regions that experienced an increase in their college-educated population, it is reasonable to hypothesize that growth in educational attainment will yield different results based on the initial level of human capital development on which it was built. Well-educated areas that work to grow their share of college-educated workers due to increasing demand may see greater benefit from such efforts than less-educated regions that promote college attendance as a way of "catching up." The literature supports both a view of well-educated workers being drawn to places with the amenities and opportunities associated with better-educated residents, as well as an understanding that areas of higher human capital levels tend to benefit more from human capital development than areas that started with lower human capital levels.

To explore the impact of existing human capital levels on regional economic outcomes, the MSAs were divided into two groups based on whether their share of the adult population with at least a bachelor's degree in 2009 was above or below the average for all regions. Dividing the MSAs based on educational attainment revealed substantial differences both in initial shares of the population with bachelor's degrees and in growth rates in adults with that level of educational attainment.

As can be seen in Table 6.1, the mean college-educated share of the population for the 179 MSAs with above-average educational attainment was nearly 14 percentage points (64 percent) greater than it was for the 216 regions with below-average educational attainment. Moreover, the change in the proportion of the regional population with a bachelor's degree or higher in the above-average educational attainment MSAs was double the rate of change for the below-average MSAs.

As would be expected given the higher shares of educational attainment in the population, the mean share of employment in occupations requiring High STEM/High Soft KSAs was higher among the above-average educational attainment MSAs than in the regions with below-average attainment. The share of regional employment in occupations requiring Low STEM/High Soft KSAs, the category with the biggest overlap with bachelor's degree requirement, was also larger for the above-average educational attainment MSAs.

However, these differences did not seem to account for the dramatic difference between the two MSA groups in terms of educational attainment. This result suggested an additional exploratory variable to be used in the model to gauge the effect of mismatch between the share of regional population with a bachelor's degree or higher compared to the share of employment in occupations requiring that level of educational attainment. As can be seen in Table 6.1, the average difference between population educational attainment and occupational educational demand was 14 percentage points among the above-average education MSAs. There was far less educational mismatch, on average, among the less-educated regions.

The apparent mismatch between regional human capital development and regional human capital deployment observed in many regions would seem to refute arguments of a widespread high-skill shortage, undercut the wisdom of so many "build it and they will come" policies, and question the utility of college completion rates as a human capital proxy. It is worth remembering that, before the formalization of human capital theory, higher education was typically assumed to be a function of consumption, not production (Schultz 1961). Higher education certainly has elements of both. People choose to pursue a college degree because of the future return it promises in the form of higher wages and better jobs (an investment in production), but they also choose to go to college for reasons such as status, family expectations, personal preference, work avoidance, and even entertainment (consumption factors). Perhaps education's duality of function as both a productive and consumptive good helps to explain the frequently mixed results from higher and increasing educational attainment that are apparent in the literature.

Table 6.1 Descriptive Statistics of the Variables for MSAs Grouped by Educational Attainment in 2009[a]

Variable	Above-average share of college graduates				Below-average share of college graduates			
	Mean	Std. dev.	Minimum	Maximum	Mean	Std. dev.	Minimum	Maximum
% Population with bachelor's or higher	34.2	6.2	26.9	58.3	20.8	3.8	11.9	26.7
% High STEM/High Soft employment	14.9	3.0	7.8	25.4	11.6	2.2	5.0	21.2
% High STEM/Low Soft employment	9.1	1.8	4.1	15.4	9.9	3.0	4.6	25.1
% Low STEM/High Soft employment	17.8	3.0	10.5	26.2	14.3	2.3	8.1	20.5
% Low STEM/Low Soft Employment	47.9	3.9	36.3	55.7	48.9	4.6	34.9	62.3
% Labor force participation	66.4	3.9	53.1	75.3	61.5	4.5	44.1	72.1
% Employment in manufacturing	9.6	3.7	2.1	23.4	12.4	6.1	2.4	36.5
Region/U.S. median house value	1.4	0.7	0.7	4.6	0.9	0.2	0.5	2.3
% Population change, 2010–2013	3.0	2.1	−1.5	9.9	1.5	2.3	−4.6	11.0
% Change in bachelor's or higher, 2009–2013	1.5	0.7	−1.7	3.6	0.8	0.8	−1.4	4.5
Percentage pt. educational mismatch	14.2	5.7	6.2	36.8	0.8	0.8	−1.4	4.5
Median wage ($)	36,398	4,945	26,720	57,430	31,326	2,902	22,780	39,190
% Change in GRP	6.7	6.3	−5.6	28.9	6.3	10.5	−9.2	70.0
Productivity ($)	112,370	25,956	67,279	214,633	93,626	19,191	63,670	215,705
Per capita income ($)	46,546	8,798	28,904	84,336	37,748	5,778	23,073	87,897
% Population below poverty line	14.1	4.1	5.5	28.0	17.2	4.0	8.2	34.8

NOTE: N for above average = 179, except for change in GRP and productivity (173); N for below average = 216, except for GRP and productivity (211) and per capita income (215).
[a] Descriptive statistics are presented without being transformed.

The descriptive statistics for the independent, dependent, and control variables are presented in Table 6.1 without being transformed for ease of interpretation. For the regression models that follow, the share of the MSA adult population with at least a bachelor's degree was recoded as a dummy variable indicating above-average and below-average educational attainment; median wage, the change in GRP, productivity, and per capita income variables were natural log transformed to correct for the skew in their distributions; and variables were standardized as z-scores.

Tables 6.2 and 6.3 present the results from a series of regression analyses testing the effects of the human capital and control variables on the five measures of regional economic well-being within the context of above-average and below-average educational attainment.

As can be seen in the two tables, the High STEM/Low Soft employment variable was the human capital variable with the broadest statistical significance in the expected direction when controlling for all other variables. Regions with a higher share of employment in High STEM/Low Soft occupations had greater growth in GRP, higher productivity, higher per capita incomes, and lower poverty rates, whether the regions had an above-average or below-average share of college-educated adults. The variable did not have a significant impact on median wages in metropolitan regions with above-average rates of educational attainment.

What is interesting is that the Low STEM/High Soft variable was significant in the expected direction on median wage, productivity, and per capita income among the 179 regions with above-average educational attainment, but it was not statistically significant on any of the measures of regional economic well-being among the 216 regions with below-average educational attainment.

High STEM/High Soft employment had a statistically significant association with all five dependent variables among the regions with below-average educational attainment, although it had a negative effect on growth in GRP.

The Low STEM/Low Soft variable was statistically significant in the expected direction on all five measures of economic well-being for the regions with below-average educational attainment, but it was only significant on two dependent variables—wages and GRP growth—for

Table 6.2 Regression Models of Occupational Skill Sets in Regions with Above-Average Share of Population with Bachelor's Degrees in 2009

Variables	Ln median wage		Ln 2013 GRP/ 2010 GRP		Ln productivity		Ln per capita income		Poverty rate	
	Coefficient	t	Coefficient	t	Coefficient	t	Coefficient	t	Coefficient	t
Intercept	0.10	1.68	-0.07	-0.88	-0.14	-1.45	-0.12	-1.51	0.06	0.64
Labor force participation	0.21	4.36***	0.21	3.39***	0.09	1.25	0.23	3.68***	-0.68	-9.74***
Manufacturing empl.	0.03	0.47	0.22	3.24***	0.06	0.75	-0.12	-1.63	0.11	1.44
Region/U.S. median house value	0.41	12.64***	0.07	1.62	0.43	8.81***	0.44	10.23***	-0.34	-7.16***
Population change, 2010–2013	-0.24	-5.75***	0.38	7.18***	0.05	0.70	-0.26	-4.58***	0.30	4.78***
% Change in bachelor's or higher, 2009–2013	-0.01	-0.28	-0.04	-0.72	0.05	0.67	-0.08	-1.38	-0.04	-0.66
Educational mismatch	-0.05	-0.85	0.04	0.50	0.04	0.44	0.31	4.15***	0.05	0.65
High STEM/High Soft employment	0.22	4.52***	-0.19	-3.03***	0.15	2.05*	0.11	1.64	-0.04	-0.51
High STEM/Low Soft employment	0.07	1.26	0.30	4.16***	0.36	4.14***	0.28	3.70***	-0.24	-2.83**
Low STEM/High Soft employment	0.19	3.40***	0.02	0.30	0.35	4.23***	0.35	4.89***	0.00	0.02
Low STEM/Low Soft employment	-0.40	-8.55***	-0.25	-4.17***	-0.08	-1.15	0.08	1.27	-0.13	-1.86
	$R^2 = 0.82$ Adj. $R^2 = 0.80$ F (df) = 73.98 (10, 168)***		$R^2 = 0.52$ Adj. $R^2 = 0.49$ F (df) = 17.52 (10, 162)***		$R^2 = 0.62$ Adj. $R^2 = 0.59$ F (df) = 26.03 (10, 162)***		$R^2 = 0.67$ Adj. $R^2 = 0.65$ F (df) = 34.38 (10, 167)***		$R^2 = 0.58$ Adj. $R^2 = 0.55$ F (df) = 22.71 (10, 168)***	

NOTE: N = 179 MSAs and NECTAs; *$p \leq 0.05$; **$p \leq 0.01$; ***$p \leq 0.001$.

Table 6.3 Regression Models of Occupational Skill Sets in Regions with Below-Average Share of Population with Bachelor's Degrees in 2009

Variables	Ln median wage		Ln 2013 GRP/ 2010 GRP		Ln productivity		Ln per capita income		Poverty rate	
	Coefficient	t	Coefficient	t	Coefficient	t	Coefficient	t	Coefficient	t
Intercept	−0.01	−0.16	−0.41	−3.03**	−0.26	−2.90**	0.02	0.25	−0.45	−4.90***
Labor force participation	0.15	4.03***	0.04	0.47	0.13	2.15*	0.24	4.71***	−0.18	−2.89**
Manufacturing empl.	0.07	2.17*	0.15	1.88	−0.07	−1.41	−0.07	−1.54	−0.14	−2.46*
Region/U.S. median house value	0.95	13.50***	−0.25	−1.51	0.30	2.73**	0.27	2.74**	−0.80	−6.90***
Population change, 2010–2013	−0.14	−3.90***	0.48	5.65***	0.10	1.74	−0.07	−1.39	0.16	2.76**
% Change in bachelor's or higher, 2009–2013	0.06	2.05*	0.02	0.33	0.03	0.54	−0.07	−1.62	−0.13	−2.66**
Educational mismatch	−0.23	−4.17***	−0.25	−1.91	−0.23	−2.61**	0.36	4.69***	−0.50	−5.37***
High STEM/High Soft employment	0.29	5.90***	−0.23	−1.99*	0.18	2.38*	0.21	3.19**	−0.21	−2.64**
High STEM/Low Soft employment	0.11	3.99***	0.33	4.96***	0.31	7.06***	0.18	4.68***	−0.20	−4.14***
Low STEM/High Soft employment	−0.03	−0.60	−0.21	−1.61	−0.15	−1.78	0.03	0.38	0.01	0.14
Low STEM/Low Soft employment	−0.26	−8.95***	−0.16	−2.29*	−0.17	−3.70***	−0.09	−2.31*	0.10	2.18*
	$R^2 = 0.68$ Adj. $R^2 = 0.67$ F (df) = 43.79 (10, 205)***		$R^2 = 0.36$ Adj. $R^2 = 0.33$ F (df) = 11.25 (10, 200)***		$R^2 = 0.44$ Adj. $R^2 = 0.41$ F (df) = 15.71 (10, 200)***		$R^2 = 0.44$ Adj. $R^2 = 0.42$ F (df) = 16.24 (10, 204)***		$R^2 = 0.49$ Adj. $R^2 = 0.47$ F (df) = 19.81 (10, 205)***	

NOTE: N = 216 MSAs and NECTAs; *$p \leq 0.05$; **$p \leq 0.01$; ***$p \leq 0.001$.

regions with an above-average share of residents with a bachelor's degree or higher.

The variable indicating change in the share of the population with at least a bachelor's degree was not significantly associated with any of the dependent variables among the regions with above-average educational attainment but was associated with higher median wages and lower poverty rates for regions with a below-average share of residents with a bachelor's degree or higher.

A higher share of college-educated residents than employment in occupations requiring a bachelor's degree or higher was associated with higher per capita incomes among both sets of regions; however, this type of educational mismatch among regions with below-average educational attainment was associated with lower median wages, lower productivity, and lower poverty rates.

What is striking in comparing the two tables is how much more the human capital models explain about the regions with above-average educational attainment. The differences in explanatory value, combined with the differences in statistical significance, tell a complex story of the effects of human capital, measured as occupational skill or educational attainment, on regional economic well-being. The benefits seem not nearly as straightforward and broad based as is frequently assumed. Context matters both in how human capital is developed and deployed.

However, one conclusion is clear: the important contribution of High STEM/Low Soft employment to regional economic well-being— at least for the period of time examined. In 2013, regions with greater shares of employment accounted for by computer programmers and geological and petroleum technicians had better economic performance than regions with a larger share of software applications developers, mathematicians, and similarly skilled workers.

What is also clear is that regions with higher shares of employment in Low STEM/Low Soft occupations suffered. They had lower or even negative GRP growth, and lower wages, productivity, and per capita incomes. Frequently, the drag on regional prosperity associated with greater employment in low-skill jobs was as large as, or larger than, the boost regions experienced from having more high-skill employment. This result suggests that regions should perhaps focus as much attention on offsetting the negative effects of low-skill work as they do on anticipating the assumed positive results of more high-skill jobs.

A MATTER OF SIZE: THE EFFECTS OF POPULATION ON HUMAN CAPITAL DEPLOYMENT

The literature indicates that there are important differences in the quality of regional human capital on the basis of population size alone. Workers in larger cities tend to be more highly educated, more highly skilled, more productive, and better paid (Glaeser and Mare 2001; Glaeser and Resseger 2010; Glaeser and Saiz 2003; Rauch 1993). Elvery (2010) found occupational skill requirements to be higher in larger regions—those with a central city population of 1 million or more—than in smaller ones.

As such, regional human capital deployment, measured in terms of occupation-based STEM and Soft KSA requirements, is expected to vary based on the size of an MSA's population. Moreover, these size-related differences in regional human capital deployment should have different effects on regional economic well-being. These expected differences present another opportunity to test the sensitivity of the occupation-based model of regional human capital.

The literature includes studies that focus on the largest MSAs, usually those with a population of 1 million or more. However, this subset was small in terms of the number of observations and did not have enough degrees of freedom to run independent tests, given the four variables of interest and five control variables. Additionally, the vast majority of MSAs continued to be grouped together below the 1 million population threshold. For the purposes of this research, the universe of MSAs was divided into thirds based on population. The group of smallest MSAs included those with populations of 65,500 or less. The biggest MSAs were those with a population of at least 176,640, with nearly half of the largest regions having populations greater than 500,000.

However, even with this effort to maintain adequate sample sizes, the substantially smaller number of study regions warranted a slight reduction in the number of variables used in the regression analyses. As such, the control variable indicating regional share of manufacturing employment, which was shown to have limited statistical significance, was eliminated from the model. The descriptive statistics presented in Table 6.4 reflect each variable's measurement before transformation for ease of discussion.

Table 6.4 Descriptive Statistics of the Variables for MSAs Grouped by Population[a]

Variable	Largest MSAs				Mid-sized MSAs				Smallest MSAs			
	Mean	Std. dev.	Min.	Max.	Mean	Std. dev.	Min.	Max.	Mean	Std. dev.	Min.	Max.
% High STEM/High Soft employment	15.5	2.8	9.9	25.4	12.9	2.2	8.0	23.1	10.8	2.1	5.0	21.2
% High STEM/Low Soft employment	9.9	1.7	6.2	16.2	9.9	3.2	5.4	25.1	8.8	2.4	4.1	19.7
% Low STEM/High Soft employment	18.7	2.4	14.3	26.2	15.5	2.2	10.0	24.2	13.5	2.2	8.1	20.1
% Low STEM/Low Soft employment	49.5	3.8	39.1	61.4	48.8	4.4	37.3	62.3	47.0	4.3	34.9	55.8
% Labor force participation	65.5	3.6	53.1	73.2	64.0	4.9	49.0	75.3	61.7	5.2	44.1	73.1
Region/U.S. median house value	1.3	0.7	0.5	4.6	1.1	0.5	0.5	3.5	0.9	0.3	0.5	2.2
% Population change, 2010–2013	2.9	2.2	-2.5	9.9	2.4	2.3	-2.4	11.0	1.2	2.3	-4.6	8.2
% Change in bachelor's or higher, 2009–2013	1.4	0.6	-0.2	2.7	1.1	0.9	-1.7	3.6	1.0	1.0	-1.4	4.5
Median wage ($)	36,131	5,189	23,540	57,430	33,111	4,027	22,780	49,340	31,616	3,548	24,530	47,220
% Change in GRP	6.4	6.7	-9.2	29.6	7.4	11.0	-7.8	70.0	5.6	8.4	-8.6	51.8
Productivity ($)	116,117	24,799	65,694	214,633	99,920	24,022	63,670	215,705	89,970	15,424	67,279	160,543
Per capita income ($)	45,270	8,510	23,073	84,336	41,799	8,955	24,802	87,897	38,140	6,344	27,483	60,304
% Population below poverty line	14.5	3.7	6.2	34.8	16.1	4.9	6.5	34.8	16.7	3.9	5.5	26.1

NOTE: $N = 132$ for the largest and mid-sized MSAs, and 131 for the smallest MSAs.

[a] Descriptive statistics are in raw data for ease of understanding; for the analysis, median wage, the change in GRP, productivity, and per capita income variables were natural log transformed, and variables were standardized.

Consistent with the literature, the larger metropolitan regions tended to have larger stocks of human capital. More than 15 percent of employment in the largest regions was in occupations requiring High STEM/High Soft KSAs, compared to less than 11 percent in the smallest MSAs. Similar differences were seen regarding the share of employment in Low STEM/High Soft occupations. Although not included in Table 6.4, educational attainment varied substantially by population size, with nearly 31 percent of adults 25 and over in large MSAs completing a bachelor's degree or higher, compared to 27 percent for mid-sized MSAs and just 23 percent for the small MSAs. Not surprisingly, the share of regional employment in occupations requiring a bachelor's degree or higher also varied by size of metropolitan region, with 21 percent of employment in the largest MSAs engaged in such occupations, compared to only 12 percent in the smallest MSAs. However, it is important to note that, for all three MSA groupings, the average population with a bachelor's degree exceeded the average share of employment in occupations requiring that level of educational attainment.

There was much less variation in the share of High STEM/Low Soft employment across the three MSA groupings, with roughly one in 10 workers, on average, engaged in such occupations regardless of regional population. Interestingly, the average share of Low STEM/Low Soft employment was also similar, although it should be noted that the OES data collection criteria capture more of the employment of larger regions than of smaller ones. Given the theorized connection between higher population and higher skill, this suggests that much of the unreported employment in the smallest regions falls into the low-skill category.

Although, on average, the largest regions had higher concentrations of human capital development and deployment, Table 6.4 demonstrates how wide those differences are. The region with the highest share of employment in occupations requiring the highest skill combination had more than 2.5 times the share of such employment in the region with the lowest. The region with the lowest share of High STEM/High Soft employment among the largest MSAs was below the average for even the least-populated regions. Moreover, among the largest regions, nearly half of employment, on average, fell into the lowest occupational skill category. This subset of highly populated regions experienced wide variations in regional economic well-being, as can be seen in the range separating the best-performing and worst-performing regions on all five dependent variables.

Tables 6.5–6.7 present the results of five separate regression analyses on the three different MSA size groupings. The model of four occupational skill and four control variables has better explanatory power for the largest MSAs than for the other two groups, especially regarding median wage (Adj. R^2 = 0.83) and per capita income (Adj. R^2 = 0.69). However, the model still explains roughly 60 percent of variation in median wage and in productivity for the smallest MSAs. The model has similar explanatory power for the largest and smallest MSAs regarding change in GRP, but much less for the mid-sized regions. The independent and control variables explain about half of the variation in poverty rates among the largest and mid-sized MSAs, but far less of the variation in poverty among the smallest regions (Adj. R^2 = 0.27).

Looking across the three tables underscores differences in the effects of how regions of varying size deploy their human capital. A higher share of employment in High STEM/High Soft occupations was associated with a higher median wage regardless of region size. However, the share of Low STEM/High Soft employment had a positive statistically significant effect on median wage, productivity, and per capita income only for the largest MSAs. The variable most associated with higher educational attainment had no statistical significance on any of the five economic well-being measures for the other two MSA population groups.

The share of High STEM/Low Soft employment was significant to regional economic well-being, but less so for the largest MSAs. A higher share of High STEM/Low Soft employment was associated with higher median wages, greater GRP growth, higher productivity, higher per capita incomes, and lower poverty among the smallest MSAs. Mid-sized regions with a higher share of High STEM/Low Soft employment saw similar benefit on all measures of regional economic well-being except for median wage. However, a higher share of High STEM/Low Soft employment was associated only with greater GRP growth and higher productivity among the largest MSAs.

As human capital theory predicts, a larger share of employment in Low STEM/Low Soft occupations was shown to be a drag on regional median wages regardless of MSA size, but the negative impact on other measures of economic well-being was seen only among the smallest regions.

Table 6.5 Relationship between STEM/Soft Occupational Skill Requirements and Regional Economic Well-Being Indicators for Largest MSAs

Variables	Ln median wage		Ln 2013 GRP/ 2010 GRP		Ln productivity		Ln per capita income		Poverty rate	
	Coefficient	t	Coefficient	t	Coefficient	t	Coefficient	t	Coefficient	t
Intercept	-0.07	-0.78	-0.11	-0.94	-0.08	-0.56	0.12	0.91	-0.16	-1.11
Labor force participation	0.21	3.40***	0.09	1.15	0.07	0.71	0.21	2.282*	-0.66	-6.70***
Region/U.S. median house value	0.40	11.39***	0.07	1.44	0.47	8.12***	0.53	10.00***	-0.29	-5.02***
Population change, 2010–2013	-0.26	-6.29***	0.29	5.28***	0.08	1.21	-0.18	-2.87***	0.25	3.70***
% Change in bachelor's or higher, 2009–2013	0.02	0.25	0.24	2.96**	-0.02	-0.16	0.10	1.10	-0.03	-0.32
High STEM/High Soft employment	0.31	4.71***	-0.17	-1.88	0.19	1.77	-0.03	-0.32	-0.01	-0.08
High STEM/Low Soft employment	0.09	1.30	0.46	5.26***	0.39	3.64***	-0.03	-0.31	-0.05	-0.43
Low STEM/High Soft employment	0.27	4.16***	-0.01	-0.10	0.36	3.27***	0.20	2.07*	0.11	1.07
Low STEM/Low Soft employment	-0.16	-2.29*	-0.16	-1.75	-0.11	-0.97	-0.16	-1.50	0.04	0.34
	$R^2 = 0.84$ Adj. $R^2 = 0.83$ F (df) = 88.89 (8, 123)***		$R^2 = 0.45$ Adj. $R^2 = 0.41$ F (df) = 12.14 (8, 120)***		$R^2 = 0.64$ Adj. $R^2 = 0.62$ F (df) = 27.10 (8, 120)***		$R^2 = 0.71$ Adj. $R^2 = 0.69$ F (df) = 37.61 (8, 123)***		$R^2 = 0.55$ Adj. $R^2 = 0.52$ F (df) = 19.03 (8, 123)***	

NOTE: N = 132 MSAs and NECTAs; *$p \leq 0.05$; **$p \leq 0.01$; ***$p \leq 0.001$.

Table 6.6 Relationship between STEM/Soft Occupational Skill Requirements and Regional Economic Well-Being Indicators for Midsized MSAs

Variables	Ln median wage		Ln 2013 GRP/2010 GRP		Ln productivity		Ln per capita income		Poverty rate	
	Coefficient	t	Coefficient	t	Coefficient	t	Coefficient	t	Coefficient	t
Intercept	0.00	-0.05	-0.02	-0.31	-0.17	-2.92**	-0.09	-2.03*	0.10	1.51
Labor force participation	0.25	6.24***	0.18	2.11*	0.08	1.21	0.24	4.92***	-0.44	-5.89***
Region/U.S. median house value	0.53	11.04***	0.01	0.13	0.28	3.72***	0.48	8.28***	-0.55	-6.12***
Population change, 2010–2013	-0.27	-6.30***	0.18	2.00*	0.03	0.50	-0.06	-1.23	0.27	3.37***
% Change in bachelor's or higher, 2009–2013	0.00	-0.11	-0.15	-1.93	-0.05	-0.89	-0.01	-0.19	-0.18	-2.56**
High STEM/High Soft employment	0.22	4.31***	-0.11	-1.04	0.12	1.55	0.13	2.14*	-0.16	-1.70
High STEM/Low Soft employment	0.07	1.94	0.34	4.41***	0.36	6.33***	0.11	2.51**	-0.29	-4.24***
Low STEM/High Soft employment	0.11	1.72	-0.13	-0.99	-0.05	-0.47	0.11	1.47	-0.01	-0.05
Low STEM/Low Soft employment	-0.33	-7.35***	-0.11	-1.17	-0.09	-1.38	-0.06	-1.07	-0.02	-0.25
	$R^2 = 0.81$ Adj. $R^2 = 0.79$ F (df) = 64.04 (8, 123)***		$R^2 = 0.32$ Adj. $R^2 = 0.27$ F (df) = 6.98 (8, 119)***		$R^2 = 0.42$ Adj. $R^2 = 0.38$ F (df) = 10.79 (8, 119)***		$R^2 = 0.61$ Adj. $R^2 = 0.58$ F (df) = 23.35 (8, 122)***		$R^2 = 0.53$ Adj. $R^2 = 0.50$ F (df) = 17.47 (8, 123)***	

NOTE: $N = 132$ MSAs and NECTAs; * $p \leq 0.05$; ** $p \leq 0.01$; *** $p \leq 0.001$.

Table 6.7 Relationship between STEM/Soft Occupational Skill Requirements and Regional Economic Well-Being Indicators for Smallest MSAs

Variables	Ln median wage		Ln 2013 GRP/ 2010 GRP		Ln productivity		Ln per capita income		Poverty rate	
	Coefficient	t	Coefficient	t	Coefficient	t	Coefficient	t	Coefficient	t
Intercept	−0.17	−2.01*	−0.28	−1.67	−0.43	−4.78***	−0.03	−0.29	−0.12	−0.99
Labor force participation	0.18	3.64***	0.14	1.43	0.05	0.87	0.12	1.91	−0.28	−3.84***
Region/U.S. median house value	0.69	7.77***	−0.28	−1.59	0.42	4.42***	0.53	4.72***	−0.37	−2.85***
Population change, 2010–2013	−0.19	−3.97***	0.46	4.86***	0.03	0.57	0.12	2.04*	0.06	0.83
% Change in bachelor's or higher, 2009–2013	0.03	0.73	0.01	0.17	−0.02	−0.41	−0.13	−2.48**	−0.05	−0.83
High STEM/High Soft employment	0.28	3.47***	−0.23	−1.44	0.00	0.05	0.02	0.23	0.01	0.06
High STEM/Low Soft employment	0.14	3.25***	0.34	4.08***	0.36	8.03***	0.22	4.21***	−0.21	−3.37***
Low STEM/High Soft employment	−0.10	−1.34	−0.22	−1.55	−0.07	−0.97	0.08	0.88	−0.01	−0.09
Low STEM/Low Soft employment	−0.35	−7.42***	−0.23	−2.44*	−0.31	−6.27***	−0.06	−1.04	−0.14	−1.96*
	$R^2 = 0.63$ Adj. $R^2 = 0.60$ F (df) = 25.84 (8, 122)***		$R^2 = 0.46$ Adj. $R^2 = 0.42$ F (df) = 12.60 (8, 118)***		$R^2 = 0.59$ Adj. $R^2 = 0.57$ F (df) = 21.44 (8, 118)***		$R^2 = 0.41$ Adj. $R^2 = 0.37$ F (df) = 10.66 (8, 122)***		$R^2 = 0.32$ Adj. $R^2 = 0.27$ F (df) = 7.06 (8, 122)***	

NOTE: N = 132 MSAs and NECTAs; *$p \leq 0.05$; **$p \leq 0.01$; ***$p \leq 0.001$.

CONCLUSION

The above analyses buttress the robustness of an occupation-based model of human capital in exploring differences in regional economic well-being. The amended models offered good explanatory power and revealed relatively consistent results regardless of whether regions were segmented by population or educational attainment levels. However, the analyses also make plain that context matters. Regions not only differ in their capacity to deploy human capital, but their success in developing their human capital may vary depending on their size and the caliber of the existing talent pool. This has important implications for policymakers from regions big and small. Policies and programs that imitate human capital investments that work in other regions without taking stock of differences in capacity and context will not likely yield the desired return in terms of improved economic well-being.

7
A Different View of the Middle

After decades of focus on "high" skills and abilities as a means of fueling the modern economy's need for "knowledge," heightened policy and media attention is being paid to "middle" skill jobs that require education and training beyond high school but less than a bachelor's degree. Headlines and program titles often include words such as *forgotten*, *overlooked*, or *vanishing* and espouse the need to "fix," "restore," or "close the gap" in middle skills.

Driving these headlines is an ongoing assertion by employers and trade and professional associations that large numbers of jobs are going unfilled because of an undersupply of workers with suitable skills. A 2011 report sponsored by the Manufacturing Institute warned that the nation was reaching a "boiling point," suggesting that as many as 600,000 jobs were going unfilled despite an era of high unemployment. Although there have been numerous articles in the academic and popular press questioning any notion of shortage, concerns over middle skills have launched public and private action. Examples include President Obama's announcement in 2015 of a $60 billion effort to provide two years of community college tuition free to qualified students, a $100 million TechHire initiative, and $175 million in apprenticeship grants made available through the Department of Labor. J.P. Morgan Chase and Co. launched a $250 million five-year "New Skills at Work" initiative to prepare workers for "high-growth, high-demand, middle-skill jobs." Moreover, many states have implemented initiatives targeting the "forgotten" middle. For example, the state of Indiana appropriated $1.2 million in 2016 toward an Apprenticeship Expansion Grant directed at graduating high school seniors and underserved populations. In addition, Iowa enacted the Apprenticeship and Training Act in 2014, tripling the amount of state funding annually appropriated for apprenticeships to $3 million, and Connecticut launched a $7.8 million Manufacturing Innovation Apprenticeship Fund in 2015.

Although the primary benefactors of such initiatives presumably are the workers who earn higher wages for their in-demand middle skills and the businesses that need appropriately skilled workers to compete, public investments in developing such skills are predicated on assumptions about the resulting public benefit—whether at the national, state, or local level. Many policies and programs directed at developing more middle-skill workers explicitly or implicitly target manufacturing and technical fields. The efforts appear to acknowledge that much of the intense policy and media focus over the past three decades on STEM has reflected a bias toward "high-skill" jobs, ignoring the importance of jobs requiring considerable technical knowledge and skill but lower levels of education.

Middle skill is largely defined by educational credentials. Jobs requiring less than a bachelor's degree but more than a high school diploma often are deemed the labor market middle. According to a report from the National Skills Coalition (2017), 53 percent of all jobs in 2015 fit such a definition. This practice of defining *skill* in terms of educational credentials or attainment, and drawing largely on entry-level requirements, obscures the wide variation in workforce demand, wages, and associated economic outcomes. The fact that so much policy attention is on reported and projected "skill shortages" in the manufacturing, technology, and health-care sectors would seem to support this observation.

This chapter presents a refined conceptualization of "middle skill" while also acknowledging the policy primacy of STEM. The research draws on the specific KSAs required of occupations to sort regional employment into "high," "middle," and "low" STEM KSAs and "high," "middle," and "low" Soft KSAs.

This method of exploring regional human capital concentrations in the context of occupational requirements for bundles of STEM and Soft KSAs provides little support for arguments advocating the importance of "middle skills," meaning the middle third of occupational skill intensity requirements, to regional economic well-being. As will be demonstrated later in this chapter, occupations requiring mid-range STEM or Soft skills contribute very little, at least for the period studied, to improving measures of regional economic well-being. What the analysis demonstrates is the importance of a relatively high level of STEM or Soft skills and the impediments to regional well-being that come from concentrations of employment in occupations requiring the lowest level of skills.

DEFINING "MIDDLE"

The literature paints two different pictures of today's labor market. One portrays job demand as an hourglass, with growth occurring at the top and bottom while jobs in the middle have been "hollowed out" (Autor, Katz, and Kearney 2008; Jaimovich and Siu 2012). The other image is of a pear, with a thicker set of jobs in the middle than at the top but with the largest girth appearing at the bottom (Holzer and Lerman 2007, 2009). Despite differing views about size of the middle, both Autor, Katz, and Kearney (2008) and Holzer and Lerman (2007) call attention to the troubling prospects of jobs becoming increasingly concentrated at the low end of the skills spectrum.

Middle-skill jobs are often defined by their wages relative to jobs paying more or less or by their educational and training requirements (Autor 2010; Autor, Katz, and Kearney 2008; Goos and Manning 2007; Holzer and Lerman 2007, 2009). Higher wages are largely assumed to reward higher levels of skill, and lower wages reflect lower skill demands. However, this would seem to ignore the effects of supply and demand. Jobs requiring relatively high skill may be relatively low-paying because a plentiful supply of candidates may be drawn to the job, owing to social prestige or psychic income. Many jobs in technical fields that require relatively low educational attainment actually pay higher wages than occupations requiring a bachelor's degree or higher because of greater demand for a limited pool of workers (Carnevale, Smith, and Strohl 2010; Carnevale, Smith, and Melton 2011).

Many of the recent political interventions focusing on middle-skill jobs appear to have arisen out of articles and reports that have sounded alarms about current or looming shortages of workers with critical skills. Such reports have been met with skepticism among scholars despite their seeming success in generating government action (Cappelli 2012, 2015; Davidson 2012; Osterman and Weaver 2014). Cappelli (2012, 2015) postulated that the shortage in technical skills actually stems from a "technical" issue: rigid software programs and keyword searches filter out many candidates who would otherwise qualify. Moreover, despite programs defining middle-skill jobs as those requiring education beyond high school but less than a bachelor's degree, Osterman and Weaver (2014) found that only 38 percent of manufacturing jobs required math skills beyond the high school level.

Holzer (2015) offered up a "tale of two middles": the traditional "middle" of good-paying construction and production jobs requiring little in terms of formal education have seen substantial declines, but the new middle includes a number of growing occupations in health care and mechanical maintenance that require higher levels of education.

Holzer (2015) also noted a rise in educational demands for traditionally low-skilled work. Educational demands that do not necessarily reflect skill requirements would seem to undercut the value of measuring middle skill as an educational middle ground between high school completion and bachelor's degree attainment. According to 2014 Occupational Employment Statistics data,[1] only 23.6 percent of total U.S. employment had a mode educational requirement of an associate's degree (8.7 percent), some college (7.6 percent), or a post–high school credential (7.4 percent). Moreover, wages for occupations requiring some college or a post–high school credential were higher, on average, than for occupations requiring only a high school diploma or less, but they were still below the national average.

ASSESSING THE MIDDLE

The differing views of middle-skill jobs, and the clearly problematic approaches to defining them, indicate a need for clarity. Given the increasing policy focus on the contribution of middle-skill work, defining such jobs in terms of skill requirements would seem to be more useful than inferring skill level based on wages and education. Moreover, the attention that business leaders, industry advocates, political leaders, and the popular press have been paying to middle-skill jobs belies a STEM priority. However, as noted previously, the academic literature provides little testing of STEM's contributions to overall economic well-being directly, let alone middle-skill STEM. A notable exception is Rothwell (2013), who identified significant differences in demand for mid-level knowledge in various STEM domains across regions.

As discussed in the previous chapters, the intense public policy focus on STEM neglects the importance of generic soft skills, despite indications from employers and evidence in the literature of their value. As with STEM, there is little in the literature that directly explores the economic effects of middle soft skills. However, the existing literature does indicate a general hypothesis:

H1. A higher share of regional employment in occupations requiring a middle level of STEM skills and a middle level of Soft KSAs is associated with positive regional economic performance.

Similar to Chapter 5, five measures of regional economic well-being were used to assess the occupation-based measures of the regional human capital asset. Regions with a larger share of employment in occupations requiring mid-level STEM and mid-level Soft KSAs are hypothesized to pay higher wages, see greater economic growth, have higher productivity, enjoy higher per capita incomes, and experience lower rates of poverty than regions with a lower share of such employment.

The method for assessing the impact of variation in concentrations of regional employment requiring mid-level STEM and mid-level Soft skills on the five measures of regional economic well-being followed the method used in Chapter 5. However, instead of labeling each of the 85 O*NET KSAs of interest as above or below average across all occupations, as described in Chapter 3, each KSA for each occupation was designated "High" if its required intensity measure (the O*NET interest and level scores combined) was in the top third, "Mid" if in the middle third, and "Low" if in the bottom third of requirements across all 942 occupations. To classify each occupation's STEM and Soft intensity, each KSA designated "High" was weighted as three points, each "Mid" KSA was weighted as two points, and each "Low" KSA was one point. Occupations were then assessed based on their collective requirements for the 35 KSAs making up the STEM bundle. The occupations with total scores in the top third of all occupations across all 35 STEM KSAs were labeled "High STEM," those in the middle third were categorized as "Mid STEM," and those in the bottom third were categorized as "Low STEM." The occupations were similarly scored on the bundle of 50 KSAs, categorizing those scoring in the top third of all occupations as "High Soft," those in the middle third "Mid Soft," and those in the bottom third "Low Soft." The STEM and Soft labels for each occupation were combined to sort all occupations into nine possible STEM/Soft categories.

As discussed in Chapter 3, the O*NET occupational data are more detailed than the information on wages and employment available from OES. Ultimately, 764 O*NET occupations with STEM/Soft designations were matched to OES employment and wage data. Table 7.1 provides the number of and share of occupations sorted into each of the nine skill categories. Appendix A presents a complete list of occupations by skill category.

Table 7.1 The Number and Distribution of Occupations Sorted by High/Mid/Low Skill Requirements

	Low STEM	Mid STEM	High STEM	Total
High Soft	53	74	95	222
	6.9%	9.7%	12.4%	29.1%
Mid Soft	98	53	92	243
	12.8%	6.9%	12.0%	31.8%
Low Soft	141	105	53	299
	18.5%	13.7%	6.9%	39.1%
Total	292	232	240	764
	38.2%	30.4%	31.4%	100.0%

SOURCE: O*NET and OES.

Table 7.2 presents the overall employment in each category, as well as each category's share of total employment covered in the OES.[2] As can be seen, just 18.5 percent of all employment was in occupations requiring High STEM skills, and only 22.7 percent of employment required High Soft skills. On the other end of the spectrum, nearly 60 percent of employment was in occupations with Low STEM requirements; 43.7 percent of employment required Low Soft skills.

Table 7.2 Employment and Distribution of Employment by High/Mid/Low Skill Requirements

	Low STEM	Mid STEM	High STEM	Total
High Soft	5,278,770	14,919,310	9,759,870	29,957,950
	4.0%	11.3%	7.4%	22.7%
Mid Soft	27,867,710	7,687,390	8,722,820	44,277,920
	21.1%	5.8%	6.6%	33.6%
Low Soft	45,030,810	6,630,800	5,915,670	57,577,280
	34.2%	5.0%	4.5%	43.7%
Total	78,177,290	29,237,500	24,398,360	131,813,150
	59.3%	22.2%	18.5%	100.0%

SOURCE: O*NET and OES.

Table 7.3 specifies the share of occupations in each of the nine STEM/Soft categories that requires a bachelor's degree or higher. As discussed in Chapter 4, overlaying the occupational educational attainment requirement on the occupational skill requirement provides support for the view of higher education as a proxy for higher skill: 83.2 percent of occupations with the highest skill requirements also required a bachelor's degree or higher. What is most interesting, however, is the high share of occupations falling into the top third in terms of Soft skill demands that require a bachelor's degree or higher—87.8 percent of the Mid STEM/High Soft and 96.2 percent of the Low STEM/High Soft occupations require a four-year college degree or more. Advanced education appears to be not nearly as necessary for occupations that demand STEM skills falling in the top third. This may partly be a reflection of the nature of the work in each category. However, as suggested in Chapter 4, this may also indicate that, for many employers, a bachelor's degree either imparts or helps to signal the presence of hard-to-assess Soft skills.

The difference in employment between the High STEM/High Soft and Low STEM/High Soft categories is also interesting. Occupations requiring a higher level of education related to STEM skills employ far fewer workers than those requiring a higher level of education related to Soft skills. Again, this may indicate differences in the nature of the

Table 7.3 Share of Occupations by High/Mid/Low Skill Category Requiring Bachelor's Degree or Higher

Skill group	No. of occupations BA+	Share of occupations BA+ (%)	Share of group employment in occupations BA+ (%)
High STEM/High Soft	79	83.2	58.6
High STEM/Mid Soft	22	23.9	24.8
High STEM/Low Soft	0	0.0	0.0
Mid STEM/High Soft	65	87.8	61.1
Mid STEM/Mid Soft	12	22.6	20.0
Mid STEM/Low Soft	1	1.0	0.0
Low STEM/High Soft	51	96.2	99.0
Low STEM/Mid Soft	31	31.6	21.1
Low STEM/Low Soft	2	1.4	0.2

SOURCE: O*NET and OES.

work, where technology-intensive activities require fewer workers than people-intensive ones.

Table 7.4 lists the variables used in this analysis, their definitions, and sources. One of the independent variables of interest—Mid STEM/Low Soft—had a moderately skewed distribution, suggesting that some form of transformation should be considered. However, a comparison of the results using the natural logarithm transformation and the untransformed variable in the regression models revealed little difference, so the untransformed variable was used to make the results easier to interpret. In addition, the nine occupational human capital variables revealed a greater potential for multicollinearity than was seen in the four-variable model in Chapter 5, but none was above a variance inflation factor of 3.1, so all of the variables were retained in the model.

RESULTS

Table 7.5 provides the mean, standard deviation, coefficient of variation, minimum, and maximum for the variables used in the regression analyses. All variables were standardized using the z-transformation to make interpretation of the results more straightforward, and the effects of the independent variables on the dependent variables are expressed in terms of standard deviations. However, the descriptive statistics reflect each variable's values before transformation, making the discussion more intuitive.

The large share of employment in low-skill occupations is immediately apparent in Table 7.5. Nearly one-third of regional employment, on average, was in occupations where KSA requirements fell in the bottom third on both the STEM and Soft dimensions. The region with the greatest concentration of Low STEM/Low Soft work had more than double the share of employment (42.9 percent) in such jobs, compared to the region with the lowest share (20.4 percent). For comparison, occupations in the top third of the STEM/Soft dimensions accounted for an average of only 3.6 percent of regional employment. However, some regions had as much as 13.2 percent of employment in these high-skill occupations.

Table 7.4 How Variables for High/Mid/Low Skill Analysis Were Defined and Calculated

Variable	Definition	Source
Dependent variables		
Median wage	Natural log of MSA median wage for all occupations	OES, May 2014
% Change in GRP	Percent change in GRP, 2009–2013	Calculated using Moody's Analytics
Productivity	MSA GRP divided by total MSA employment, 2013	Calculated using Moody's Analytics
Per capita income	MSA per capita income for the previous 12 months in 2013 $	ACS 5-year estimate, 2013
Poverty rate	Share of MSA population below the poverty line, 2013	ACS 5-year estimate, 2013
Independent variables		
High STEM/High Soft employment	Share of MSA employment in occupations requiring both top 33% STEM and top 33% Soft skills	Calculated using O*NET 19.0 and OES, May 2014
High STEM/Mid Soft employment	Share of MSA employment in occupations requiring top 33% STEM skills but middle 33% Soft skills	Calculated using O*NET 19.0 and OES, May 2014
High STEM/Low Soft employment	Share of MSA employment in occupations requiring top 33% STEM skills but bottom 33% Soft skills	Calculated using O*NET 19.0 and OES, May 2014
Mid STEM/High Soft employment	Share of MSA employment in occupations requiring middle 33% STEM but top 33% Soft skills	Calculated using O*NET 19.0 and OES, May 2014
Mid STEM/Mid Soft employment	Share of MSA employment in occupations requiring both middle 33% STEM skills and middle 33% Soft skills	Calculated using O*NET 19.0 and OES, May 2014
Mid STEM/Low Soft employment	Share of MSA employment in occupations requiring middle 33% STEM skills but bottom 33% Soft skills	Calculated using O*NET 19.0 and OES, May 2014

(continued)

Table 7.4 (continued)

Variable	Definition	Source
Low STEM/High Soft employment	Share of MSA employment in occupations requiring bottom 33% STEM but top 33% Soft skills	Calculated using O*NET 19.0 and OES, May 2014
Low STEM/Mid Soft employment	Share of MSA employment in occupations requiring bottom 33% STEM skills but middle 33% Soft skills	Calculated using O*NET 19.0 and OES, May 2014
Low STEM/Low Soft employment	Share of MSA employment in occupations requiring both bottom 33% STEM skills and bottom 33% Soft skills	Calculated using O*NET 19.0 and OES, May 2014
Control variables		
Labor force participation	Share of the MSA population age ≥ 16 in the labor force, 2013	Calculated using ACS 5-year estimate, 2013
Manufacturing employment	Share of the MSA total employment in manufacturing	Calculated using ACS 5-year estimate, 2013
Region/U.S. median house value	Regional median house value/U.S. median house value	Calculated using ACS 5-year estimate, 2013
Population change, 2010–2013	Percent change in MSA population 2010–2013	Calculated using ACS 5-year estimate, 2013
% Change in bachelor's or higher, 2009–2013	Percent change in share of MSA population age ≥ 25 with a bachelor's degree or higher, 2009–2013	Calculated using ACS 5-year estimate, 2013
Education mismatch	Difference between share of MSA population with a bachelor's degree or higher and share of occupational employment requiring a bachelor's degree or higher	Calculated using ACS 5-year estimate, 2013 and OES data, May 2014

Table 7.5 Descriptive Statistics for Variables Used in High/Mid/Low Skill Analysis[a]

Variable	Mean	Std. dev.	CV	Min.	Max.
% High STEM/High Soft employment	3.6	1.7	0.5	0.7	13.2
% High STEM/Mid Soft employment	4.4	1.2	0.3	1.7	9.7
% High STEM/Low Soft employment	4.0	1.1	0.3	1.5	10.4
% Mid STEM/High Soft employment	12.1	1.9	0.2	5.8	17.8
% Mid STEM/Mid Soft employment	5.8	1.4	0.2	2.4	13.1
% Mid STEM/Low Soft employment	5.9	2.4	0.4	1.4	22.3
% Low STEM/High Soft employment	2.6	1.1	0.4	0.6	7.2
% Low STEM/Mid Soft employment	18.6	2.4	0.1	11.1	25.4
% Low STEM/Low Soft employment	29.9	3.2	0.1	20.4	42.9
% Labor force participation	63.7	4.9	0.1	44.1	75.3
% Employment in manufacturing	11.1	5.3	0.5	2.1	36.5
Region/U.S. median house value	1.1	0.6	0.5	0.5	4.6
% Population change, 2010–2013	2.2	2.4	1.1	−4.6	11.0
% Change in bachelor's or higher, 2009–2013	1.2	0.9	0.8	−1.7	4.5
Education mismatch (% pt.)	10.6	5.7	0.5	−1.2	36.8
Median wage	$33,624	$4,694	0.1	$22,780	$57,430
% Change in GRP	6.5	8.9	1.4	−9.2	70.0
Productivity ($)	$102,071	$24,325	0.2	$63,670	$215,705
Per capita income ($)	$41,745	$8,514	0.2	$23,073	$87,897
% Population below poverty line	15.8	4.3	0.3	5.5	34.8

NOTE: $N = 395$, except for GRP and productivity (384) and per capita income (394).

[a] Descriptive statistics are in raw data for ease of understanding; for the analysis, the wage variable was log transformed, and all variables were standardized using the z-transformation.

Mid-Level Skills Have Limited Effect on Regional Economic Well-Being

Table 7.6 provides the results of regression analyses on the five dependent variables that constitute different aspects of regional economic well-being. Compared to the four-category occupational skill models explored in Chapter 5, little to no improvement in explanatory value can be seen for four of the economic well-being measures for the revised models using the nine-category occupational human capital variables and the two education control variables. On the poverty measure, however, the revised model increased explanatory value considerably (Adj. $R^2 = 0.53$, compared to Adj. $R^2 = 0.46$).

Only two of the occupation-based measures were statistically significant across all five economic well-being measures—High STEM/ Low Soft and Low STEM/Low Soft. Consistent with the previous models, regions with larger shares of employment in occupations with skill requirements that put them in the top third across all occupations for STEM and the bottom third for Soft KSAs had higher median wages, greater growth in GRP, higher productivity, higher per capita incomes, and lower poverty rates. Conversely, regions with a larger share of employment requiring STEM and Soft KSAs ranking in the bottom third across all occupations had lower median wages, lower growth in GRP, lower productivity, lower per capita incomes, and higher poverty rates.

By definition, the Mid STEM/Mid Soft variable represents the middle in terms of occupational skill requirements. However, the impact of these middle skills is limited to only one measure of regional economic well-being. Regions with a higher share of Mid STEM/Mid Soft employment had higher productivity, controlling for all other variables. A middle level of STEM skills, in combination with higher or lower Soft skills, also had limited effect. Regions with a higher share of Mid STEM/High Soft employment had higher per capita incomes. Regions with a higher share of employment in occupations requiring Mid STEM skills but Low Soft skills had lower median wages and higher poverty rates.

A higher share of employment requiring Low STEM/Mid Soft KSAs was associated with lower median wages, lower GRP growth, and lower productivity. However, regions with a higher share of Low

Table 7.6 Relationship between High/Mid/Low Occupational Skill Requirements and Regional Economic Well-Being Indicators

Variables	Ln median wage Coefficient	t	% Change in GRP Coefficient	t	Productivity Coefficient	t	Per capita income Coefficient	t	Poverty rate Coefficient	t
Intercept	0.00	3795***	0.01	0.11	0.01	0.26	0.01	0.32	0.00	0.00
Labor force participation	0.19	6.57***	0.14	2.48*	0.14	3.09**	0.19	4.63***	−0.39	−7.89***
Manufacturing employment	0.00	−0.01	0.13	2.47*	−0.04	−0.90	−0.05	−1.30	−0.08	−1.77
Region/U.S. median house value	0.44	15.5***	0.04	0.74	0.47	10.78***	0.50	12.73***	−0.42	−8.74***
Population change, 2010–2013	−0.14	−4.84***	0.42	7.48***	0.03	0.71	−0.09	−2.14*	0.19	3.90***
% Change in bachelor's or higher, 2009–2013	0.03	1.06	−0.01	−0.24	0.04	1.10	−0.06	−1.85	−0.12	−2.86**
Educational mismatch	−0.05	−1.60	−0.05	−0.82	−0.14	−2.94**	0.15	3.61***	0.05	0.90
High STEM/High Soft employment	0.28	7.68***	−0.18	−2.56**	0.05	0.90	0.09	1.88	−0.08	−1.28
High STEM/Mid Soft employment	0.01	0.28	0.16	2.39*	0.25	4.53***	0.04	0.83	0.02	0.41
High STEM/Low Soft employment	0.15	4.79***	0.24	4.19***	0.18	3.81***	0.23	5.40***	−0.26	−5.01***
Mid STEM/High Soft employment	−0.01	−0.26	−0.01	−0.21	0.04	0.88	0.10	2.33*	0.07	1.39
Mid STEM/Mid Soft employment	−0.01	−0.38	−0.03	−0.45	0.13	2.37*	0.03	0.57	−0.02	−0.33
Mid STEM/Low Soft employment	−0.11	−4.24***	−0.01	−0.11	−0.02	−0.44	−0.04	−1.17	0.10	2.19*
Low STEM/High Soft employment	0.31	8.66***	0.08	1.22	0.08	1.39	0.04	0.72	0.12	2.06*
Low STEM/Mid Soft employment	−0.15	−4.94***	−0.25	−4.18***	−0.15	−3.13**	0.06	1.54	−0.26	−5.13***
Low STEM/Low Soft employment	−0.22	−8.97***	−0.14	−2.98**	−0.13	−3.36***	−0.07	−1.99*	0.11	2.61**
	$R^2 = 0.84$		$R^2 = 0.41$		$R^2 = 0.61$		$R^2 = 0.67$		$R^2 = 0.55$	
	Adj. $R^2 = 0.83$		Adj. $R^2 = 0.39$		Adj. $R^2 = 0.60$		Adj. $R^2 = 0.66$		Adj. $R^2 = 0.53$	
	$F(df) = 129.52$		$F(df) = 17.24$		$F(df) = 39.02$		$F(df) = 52.13$		$F(df) = 30.21$	
	$(15, 379)$***		$(15, 368)$***		$(15, 368)$***		$(15, 378)$***		$(15, 379)$***	

NOTE: $N = 395$ MSAs and NECTAs; *$p \leq 0.05$; **$p \leq 0.01$; ***$p \leq 0.001$.

STEM/Mid Soft employment also tended to have lower poverty rates. Regions with a higher share of employment requiring High STEM/Mid Soft KSAs had greater growth in GRP and higher productivity.

The two remaining occupation-based measures were associated with higher median wages, but each also had a negative association with one of the measures of regional economic well-being. Regions with a larger share of High STEM/High Soft employment tended to experience lower growth in GRP, while regions with more employment accounted for by Low STEM/High Soft occupations tended to have higher poverty rates.

The two human capital measures tied to regional educational attainment levels both had limited statistical significance. An increasing share of college-educated adults in a region's population was associated with lower poverty rates, whereas regions with higher levels of educational mismatch (the percentage of adults with high educational attainment minus the percent of occupational employment that requires at least a bachelor's degree) is associated with lower productivity levels but higher per capita incomes. This is a rather perplexing result. Higher per capita incomes may be explained by better-educated workers requiring higher pay, even for occupations that do not necessarily require a high level of education. The lower productivity finding, however, seems to call into question the assumption that higher levels of education make workers more productive or suggests that these workers are disproportionately employed in service sector occupations with relatively low levels of productivity.

Sorting Regions by Skill Concentrations

A complete list of the occupations where STEM KSA requirements were in the top third and Soft skill requirements were in the bottom third across all occupations is included in Appendix A. It is useful to explore the occupations that were grouped into this category. The largest number of jobs among the High STEM/Low Soft occupations was for general maintenance and repair workers, who accounted for nearly 1.3 million workers nationwide. The High STEM/Low Soft category also includes industrial machinery mechanics, machinists, construction laborers, tool and die makers, pattern makers, pipefitters, electro-mechanical equipment assemblers, solar photovoltaic installers, oil and gas derrick operators, and oil and gas roustabouts.

Table 7.7 lists the 14 MSAs with the largest share of employment in High STEM/Low Soft occupations. Among the top regions were three MSAs from Texas and three from Louisiana. The impact of the oil and gas industry is clear, with prices per barrel topping $100 in 2013. The Texas MSAs encompassing Odessa and Midland both sit atop the Permian Basin, which accounts for more than 23 percent of U.S. crude oil production and which the U.S. Geological Survey, in a November 2016 report, estimated as having up to an additional 20 billion barrels of recoverable oil due to changing technology and industry practices. Beaumont-Port Arthur, Texas, is home to the nation's largest oil refinery. The economies of Houma-Bayou Cane-Thibodaux, Louisiana;

Table 7.7 Top Regions for High STEM/Low Soft and High STEM/Mid Soft Employment

High STEM/Low Soft employment		High STEM/Mid Soft employment	
MSA	Employment share (%)	MSA	Employment share (%)
Odessa, TX	10.4	Midland, TX	9.7
Houma-Bayou Cane-Thibodaux, LA	8.7	Odessa, TX	9.7
Midland, TX	8.6	Beaumont-Port Arthur, TX	9.1
Farmington, NM	8.3	Casper, WY	8.0
Casper, WY	8.1	Corpus Christi, TX	7.5
Lake Charles, LA	7.7	Warner Robins, GA	7.5
Greeley, CO	7.7	Houma-Bayou Cane-Thibodaux, LA	7.5
Lafayette, LA	7.4	Framingham, MA	7.4
Elkhart-Goshen, IN	6.9	Farmington, NM	7.3
Beaumont-Port Arthur, TX	6.7	Greeley, CO	7.3
Sheboygan, WI	6.7	Bremerton-Silverdale, WA	7.2
Corpus Christi, TX	6.6	Baton Rouge, LA	7.0
Williamsport, PA	6.5	Burlington, NC	7.0
Rockford, IL	6.4	Houston-Sugar Land-Baytown, TX	6.8
		Lake Charles, LA	6.7

SOURCE: O*NET and OES.

Farmington, New Mexico; Casper, Wyoming; Lake Charles, Louisiana; and Greeley, Colorado, all have been powered by the energy industry. From 2009 to 2013, GRP for these 14 metropolitan regions grew by an average of more than 25 percent, propelled by a doubling of oil prices. The average productivity level for 2013 across these 14 regions was nearly $133,000, or some $30,000 above the mean for all MSAs. Per capita income across the 14 regions was nearly 8 percent above the average for all MSAs, but the median wage for the 14 High STEM/Low Soft regions was only 1 percent above the overall average.

As can be seen in Table 7.7, there is considerable overlap between the regions with the highest share of employment in High STEM/Low Soft occupations and the other occupation-based human capital measure associated with greater growth in GRP and higher productivity, High STEM/Mid Soft. Nine of the 15 regions top the lists for both categories. The High STEM/Mid Soft category includes occupations related to the oil and gas industry, such as geological and petroleum technicians; rotary drill operators; petroleum pump system operators; and crushing, grinding and polishing machine setters; as well as chemical technicians and chemists. The category also includes a variety of technical workers, such as electro-mechanical technicians, industrial machinery technicians, computer numerically controlled (CNC) machine tool operators and programmers, medical equipment repairers, nuclear medicine technologists, and health technologists and technicians. In addition, several computer-related occupations are categorized as High STEM/Mid Soft, such as software application developers, computer hardware engineers, and computer user support specialists.

Certainly, the broad benefits to regions from higher concentrations of High STEM/Low Soft employment and the GRP and productivity gains associated with larger shares of High STEM/Mid Soft employment have been fueled, at least in part, by a boom in the oil and gas industry enabled by changes in technology and industrial processes. The bust period in the industry that took hold in the latter half of 2014 would likely reveal a negative impact on the very regions that were experiencing the benefits of a surging industry just a few years earlier. The oil and gas business and its related upstream processing industries are notoriously cyclical, with boom-bust cycles being driven by the real cost of both oil and natural gas. Although most industries do not experience the rapid-cycle highs and lows seen in the oil and gas industry over the past four decades, the results underscore the fact that regional human capital stocks and regional economic well-being largely reflect the health of the region's portfolio of industries. This is especially true of regions dominated by one or two industries.

So much of the policy focus and activities regarding STEM has been on increasing the supply of workers with STEM skills, assuming that a larger pool of well-trained talent will energize local and state economies. Yet, such efforts are bounded by geographic demand for that human capital. The interplay presents a challenge for human-capital-based policies that must balance support for local industries' efforts in addressing immediate workforce needs while encouraging the development of human capital suitable for future opportunities.

Table 7.8 lists the regions with the highest share of employment in the occupation-based human capital category that best captures what is meant when political leaders, pundits, and educators talk about STEM jobs. This category encompasses various engineers, computer and information research scientists, computer systems analysts, physicists, biochemists and biophysicists, materials scientists, various doctors, and medical scientists, but it also captures business operations specialists, industrial production managers, science teachers, and nurse practitioners. As would be expected, a sizable share of workers in familiar West Coast information technology hubs—Silicon Valley, the San Francisco Bay area, and Seattle—are engaged in High STEM/High Soft activities. Boulder, Colorado, has also emerged as a national leader in tech startup activity (Dill 2015). As two of the three hubs making up North Carolina's renowned Research Triangle Park, Durham-Chapel Hill is dominated by two research universities, pharmaceutical and other life science organizations, and information technology businesses.

In addition to information technology and biotech, the list of MSAs in Table 7.8 highlights the demand for High STEM talent related to the activities of the federal government. A Brookings Institution report ranked Washington, D.C., as third in the nation for job postings requiring STEM skills (Rothwell 2013). Huntsville, Alabama; Warner Robins, Georgia; San Diego; and Denver are all hubs for aerospace and defense activities, in part serving the needs of the military, NASA, and homeland security. For 60 years, Warren, Michigan, has been the location for innovation into ground systems as part of the U.S. Army's Tank Automotive Research, Development, and Engineering Center.

Table 7.8   Top Regions for High STEM/High Soft and Low STEM/High Soft Employment

High STEM/High Soft employment		Low STEM/High Soft employment	
MSA	Employment share (%)	MSA	Employment share (%)
Huntsville, AL	13.2	New York-White Plains-Wayne, NY-NJ	7.2
Warner Robins, GA	12.3	Boston-Cambridge-Quincy, MA	7.0
San Jose-Sunnyvale-Santa Clara, CA	9.9	Washington-Arlington-Alexandria, DC-VA-MD-WV	6.5
Boulder, CO	8.7	Bridgeport-Stamford-Norwalk, CT	6.3
Washington-Arlington-Alexandria, DC-VA-MD-WV	8.2	San Francisco-San Mateo-Redwood City, CA	6.2
Framingham, MA	7.9	Ithaca, NY	6.1
Bethesda-Rockville-Frederick, MD	7.9	Philadelphia, PA	5.6
Durham-Chapel Hill, NC	7.8	Hartford-West Hartford-East Hartford, CT	5.4
Seattle-Bellevue-Everett, WA	7.7	Wilmington, DE-MD-NJ	5.3
Sacramento-Arden-Arcade-Roseville, CA	7.4	Albany-Schenectady-Troy, NY	5.2
Detroit-Livonia-Dearborn, MI	7.4	Worcester, MA-CT	5.2
Trenton-Ewing, NJ	7.3	Des Moines-West Des Moines, IA	4.9
San Francisco-San Mateo-Redwood City, CA	7.2	Springfield, MA-CT	4.8
Warren-Troy-Farmington Hills, MI	7.2	Providence-Fall River-Warwick, RI-MA	4.8
Denver-Aurora-Broomfield, CO	7.1	Burlington-South Burlington, VT	4.7

SOURCE: O*NET and OES.

Bethesda, Maryland, is home to the National Institutes of Health, as well as the Walter Reed National Military Medical Center. Markusen (1984) and Flamm (1988), among others, called attention to the connection between government spending and the rise of high-tech centers. In addition to direct defense spending, many of the regions with a large share of High STEM/High Soft employment also are anchored by research universities receiving state and federal support.

Given the tendency to link STEM skills to new, disruptive technologies, it might seem surprising that Detroit, the hub of disruptive innovation a century ago, continues to be a top consumer of High STEM/High Soft talent. But the automotive industry has shifted its emphasis in southeastern Michigan from a dominance of assembly activities to research and development, which relies on High STEM/High Soft skills.

Two regions with high shares of employment in High STEM/High Soft occupations—Washington, D.C., and San Francisco—are also among the top regions for Low STEM/High Soft employment. Both occupation-based human capital measures were associated with higher regional median wages. The number of regions that are either seats of government or anchored by research universities, or both, stands out in Table 7.8. This finding is not surprising, given that the Low STEM/High Soft category was most associated with higher educational requirements. All 15 regions in Table 7.8 in the Low STEM/High Soft category had above-average educational attainment, and all but two had populations ranking in the top third among all MSAs. This suggests a strong connection between region size, higher education, and High Soft skills, even more so than High STEM skills. Aside from activities associated with government and education services, many of the regions with the largest share of Low STEM/High Soft employment also are dominated by corporate headquarters, research and product development facilities, and financial services and insurance companies.

Table 7.9 lists the regions with the highest share of employment in the two categories of occupation-based human capital associated with a drag on most or all measures of regional economic well-being—Low STEM/Mid Soft and Low STEM/Low Soft. What stands out on both lists of regions is the prevalence of metropolitan areas reliant on tourism, health care, and low-value or declining manufacturing. Two regions with the largest share of low-skill employment—New York-White

Table 7.9 Top Regions for Low STEM/Mid Soft and Low STEM/Low Soft Employment

Low STEM/Mid Soft employment		Low STEM/Low Soft employment	
MSA	Employment share (%)	MSA	Employment share (%)
Fort Lauderdale-Pompano Beach-Deerfield Beach, FL	25.4	Las Vegas-Paradise, NV	42.9
Sioux Falls, SD	25.3	Brownsville-Harlingen, TX	39.6
Miami-Miami Beach-Kendall, FL	25.3	Atlantic City-Hammonton, NJ	39.4
New York-White Plains-Wayne, NY-NJ	24.8	Myrtle Beach-North Myrtle Beach-Conway, SC	38.8
Rapid City, SD	24.5	McAllen-Edinburg-Mission, TX	37.3
Tampa-St. Petersburg-Clearwater, FL	24.2	Laredo, TX	37.0
Nassau-Suffolk, NY	24.2	Lebanon, PA	36.6
Salt Lake City, UT	23.9	Ocean City, NJ	36.4
Peabody, MA	23.0	Flint, MI	36.3
Raleigh-Cary, NC	22.9	Elkhart-Goshen, IN	36.2
Tallahassee, FL	22.8	Asheville, NC	36.0
Jacksonville, FL	22.8	Riverside-San Bernardino-Ontario, CA	36.0
West Palm Beach-Boca Raton-Boynton Beach, FL	22.8	Youngstown-Warren-Boardman, OH-PA	36.0
Jackson, MS	22.7	Reno-Sparks, NV	35.7
Denver-Aurora-Broomfield, CO	22.7		

SOURCE: O*NET and OES.

Plains-Wayne and Denver-Aurora-Broomfield—are also among the regions with the largest share of high-skill or high-education employment, and Raleigh-Cary is the third hub of North Carolina's Research Triangle. The high share of employment concentrated in the lowest skill category is striking. Despite nine possible skill categories, 43 percent of Las Vegas employment was in Low STEM/Low Soft occupations. All 14 regions in the Low STEM/Low Soft category in Table 7.9 had more than 35 percent of employment in occupations requiring the lowest combination of skill.

CONCLUSION

Although a great deal of policy attention is devoted to supporting High STEM college majors and jobs (a focus largely supported by the results presented here), High Soft skills are also associated with measures of regional economic well-being, particularly higher median wages. In combination with Mid STEM skills specifically, High Soft skills were linked to higher per capita incomes, which suggests that High Soft skills may be more important to the well-being of individuals in a region (in terms of wages and per capita incomes) than to the economic competitiveness of the region itself. Mid STEM/High Soft occupations include chief executives, sales managers, human resource managers, financial specialists, and various medical and health care professionals.

Occupations requiring High STEM skills, in general, appear to make a difference in regional economic performance. These findings would seem to support the considerable attention paid to STEM skills by government leaders. However, the focus may be somewhat misplaced. Although 83 percent of the 95 occupations requiring both High STEM and High Soft skills also require a bachelor's degree or more, only a quarter of High STEM/Mid Soft occupations and no High STEM/Low Soft occupations require such high levels of educational attainment. High STEM/Mid Soft and High STEM/Low Soft were the only occupation-based human capital measures associated with higher rates of GRP growth and higher productivity. High STEM/Low Soft employment was the only variable exhibiting the theorized and desired human capital effect across all five measures of regional economic well-being.

In addition, regional employment in occupations requiring High STEM/Mid Soft skills had the desired effect on more measures of regional economic health than did employment in High STEM/High Soft occupations, which are the focus of much of the policy and rhetoric about the importance of STEM. This would suggest that policies are overlooking paths to connect workers to High STEM jobs by focusing too intently on educational attainment. Many occupations requiring a relatively high level of STEM skill require relatively low levels of formal education.

This finding largely bolsters arguments made by Holzer (2008) and Rothwell (2013) suggesting a higher education bias in STEM policy and conceptualization. However, this research offers little support for assertions that middle STEM skills—defined here as those falling in the middle third of occupational requirements across 35 individual STEM-related KSAs—are important contributors to regional economic performance. As noted earlier, there is a positive association between having larger shares of Mid STEM/High Soft employment and per capita incomes. A positive association also exists between the proportion of a region's employment in Mid STEM/Mid Soft occupations and GRP growth rate. These findings demonstrate that mid-level STEM skills do play a role, although a narrow one, in the economic vitality of regions. However, the regional benefits of mid-level STEM KSAs were seen only when combined with at least mid-level Soft skills. Regions with a higher share of employment in occupations requiring Mid STEM/Low Soft skills actually had lower median wages and higher poverty rates.

Again, although much of the recent advocacy and interventions targeted toward training workers for middle-skill jobs tend to zero in on STEM-related technical and mechanical activities, these results indicate that greater support for mid-level Soft skills are warranted. Regions with higher shares of employment in occupations requiring Mid Soft skills in combination with at least Mid STEM KSAs tended to be places of higher productivity and, to a certain extent, greater economic growth.

The demonstrated narrow significance of employment concentrations in occupations labeled "Mid" for this analysis should not necessarily undercut the importance of middle-skill jobs to individual workers as well as to regional economies. Presumably, occupations that require neither the highest levels of skill (High STEM/High Soft) nor the lowest (Low STEM/Low Soft) can be thought of as "middle." Additionally, the limited effect appears largely to be one of definition. A third of the occupations categorized as High STEM for this analysis—and 42 percent of High STEM employment—would be categorized as "middle skill" based on the educational criterion of requiring more than a high school diploma but less than a bachelor's degree.

Assuming the 62 percent of U.S. employment in occupations not captured by the highest or lowest STEM/Soft categories constitutes what is meant by middle-skill jobs, policymakers are right to expand their focus on human capital development beyond simply increasing college going. Regions with larger shares of employment in occupations requiring High STEM skills in combination with Mid Soft or Low

Soft KSAs witnessed better economic performance as the nation was climbing out of the Great Recession—the time period covered by this study.

Muddying policy efforts, however, is an apparent conflict between what is good for a region—or state or nation—overall and the possible return for individuals. Figure 7.1 shows how median wages, measured across all occupations at the national level, vary by skill requirements. The importance of superior Soft skills to worker wages is apparent. Occupations falling in the top third in terms of Soft skill requirements pay substantially more than all other occupational skill categories. Workers employed in occupations requiring Mid STEM but High Soft KSAs earn far more than workers in High STEM/Mid Soft occupations. What is also apparent is how little workers in High STEM/Low Soft occupations are rewarded for the economic benefit they may be returning to regions. Workers in High STEM/Low Soft occupations were paid some of the lowest wages, even though regions with a higher share of such workers enjoyed broad-based economic benefits.

One last observation should be drawn from a comparison of the findings across the previous chapters: that is, the apparent link among industrial demand, occupational skill requirements, and regional economic well-being. As is largely assumed in the literature, the media, and policy initiatives, many regions that are experiencing greater economic

Figure 7.1 Median Wage ($) by STEM/Soft Category

	Low STEM	Mid STEM	High STEM
High Soft	61,450	72,845	79,930
Mid Soft	41,745	46,690	50,785
Low Soft	26,640	35,420	39,100

well-being, at least on some of the measures explored, have higher concentrations of employment in occupations such as software application developers and computer network architects. However, many regions that enjoyed greater economic well-being across all five measures had higher concentrations of employment in occupations related to the oil and gas industry and other occupations in industries supporting oil and gas activity. The time frame of this analysis reflected a period during which technological innovations and world energy prices fueled an economic boom in the U.S. oil and gas industry. This observation underscores how intertwined human capital demand is with industrial demand. It also highlights the challenge of identifying specific skill sets for policy support.

Notes

1. Available at https://www.bls.gov/oes/ (accessed June 29, 2017).
2. As noted in Chapter 5, the OES survey does not include self-employed workers, and certain occupations are not included in this analysis. The 764 occupations captured, on average, 86.9 percent of regional employment.

8
Cross-Cutting Skills

The previous chapters have demonstrated wide variation in how human capital is deployed across regions and the effects of that variation on regional economic well-being. However, the goal of policies and interventions to encourage more people to attend college, and more specifically to study STEM subjects, is ensuring that regions have the appropriate supply of valuable human capital to meet changing demands for talent. Peering beneath the STEM/Soft category labels to explore specific skills, knowledge domains, and abilities shared across occupations offers the opportunity to better align these efforts to improve the regional supply of in-demand human capital.

In many respects, the attention directed at expanding STEM education contributes to a narrow understanding of the knowledge and skills valued and deployed in the workplace. The focus on specific STEM knowledge domains itself downplays the broad application of Soft skills. Breadth of utilization, in fact, would seem to be an important factor guiding both public and private decision making. Insight into potentially overlapping skill sets would provide a good foundation for guiding education and economic development policies, as well as personal and business choices regarding human capital investments.

"High" KSAs that cut across occupations within specific industries or across multiple industries represent a critical asset that regional policy could help to support or build. Workers possessing a high level of such critical skills would benefit from more extensive job prospects within the region—and presumably higher pay; workers in occupations requiring a middle level of such KSAs would know what human capital to develop and highlight to enhance their job prospects and increase their earnings. Employers would benefit from knowing the depth and breadth of pools of talent from which they could draw.

A REVISED METHOD

Attempts to break down jobs and occupations into their component parts and give concrete information on the building blocks for career success have been gaining traction nationwide and at all levels of government. For example, the integrative Career Pathways workforce development strategy helps middle school and high school students match their interests and abilities to possible careers. It also indicates which skills may help workers transition into new but related occupations.

The previous chapters demonstrated the value of larger concentrations of "high" skills to regions. But are certain "high" skills shared broadly across each of the nine STEM/Soft categories? And, more generally, what skills characterize different human capital concentrations?

The analysis presented here offers a slightly different view of "high" skill than is evident in the current public policy discussion. Instead of assuming a ranking in which some KSAs are "higher" than others, this analysis has taken the approach that each and every KSA may be required at a "high" level in certain settings. In other words, occupations were assessed on how high or low their requirements were for each KSA making up the STEM and Soft skill bundles.

This examination involves a slight departure from the method described in Chapter 3 and previously followed to sort occupations into the various skill categories. Whereas the skill intensity score used to categorize occupations incorporated the two dimensions of interest and level that O*NET provides to characterize occupational demand for each KSA, only the level score is used here to explore "high" and "low" skills. Also, the number of KSAs examined was reduced from 85 to 67, in an effort to align the KSAs with the bundles of baseline and STEM-related skills put forth by the job market analytics firm Burning Glass Technologies.

Each occupation was assessed on how its required level of knowledge or skill compared to the mean for each KSA. Occupations were assumed to require "high" or "very high" skill if their level scores were one or two standard deviations above the mean, respectively. Conversely, occupations were labeled "low" or "very low" if their typical level of requirement was one or two standard deviations below the mean. The primary interest of this examination is high skills that cut across occupations in each of the nine human capital categories. However, awareness of low skill requirements was helpful in characterizing each category.

Interestingly, nearly 90 percent of occupations required at least a high level (i.e., at least one standard deviation above the mean) on at least one of the STEM or Soft KSAs. Thus, the vast majority of occupations can be thought of as requiring "high" skill to some extent.

Other insights also came to light by characterizing occupational skill requirements in this manner. Take, for example, the O*NET knowledge domain of engineering and technology, which presumably by all standards would be expected to represent STEM activities. Nearly 10 percent of all occupations were found to require a very high or high level of engineering and technology knowledge. Occupations requiring a very high level of engineering and technology knowledge included, as would be expected, various engineering occupations, as well as physicists, material scientists, and commercial and industrial designers. Other, perhaps less obvious, occupations requiring a very high level of such engineering and technology knowledge included mechanical drafters, electronic and electrical engineering technicians, and electrical and electronics repairers of commercial and industrial equipment. As also would be expected, most of these occupations requiring a very high level of engineering and technology knowledge required a bachelor's degree or higher, but several did not.

Occupations requiring only a high level of engineering and technology knowledge included computer and information research scientists, software developers, computer network support specialists, information security analysts, industrial production managers, and CNC machine tool programmers. Again, many of these occupations required a bachelor's degree or higher, but many did not.

RESULTS

Table 8.1 provides the KSAs that are shared across the most occupations for each human capital category. The table indicates that High STEM/High Soft occupations tended to require a very high level (at

Table 8.1   KSAs Most Shared by Occupations in Each Human Capital Category

"Very high" KSAs	Share of occupations (%)	"High" KSAs	Share of occupations (%)	"Low" or "very low" KSAs	Share of occupations (%)
High STEM/High Soft occupations					
Physics	26.3	Complex problem solving	53.7	Equipment selection	12.6
Biology	24.2	Judgment and decision making	53.7	Operation and control	12.6
Engineering and technology	22.1	Monitoring	51.6	Quality control analysis	6.3
Design	22.1	Systems analysis	48.4	Troubleshooting	6.3
Science	21.1	Reading comprehension	47.4	Operation monitoring	5.3
Technology design	20.0	Science	46.3	Selective attention	4.2
Problem sensitivity	18.9	Operations analysis	46.3	Perceptual speed	3.2
		Information ordering	46.3	Time sharing	3.2
High STEM/Mid Soft occupations					
Repairing	17.2	Equipment selection	57.5	Operations analysis	11.5
Equipment maintenance	13.8	Equipment maintenance	50.6	Service orientation	8.0
Telecommunications	13.8	Troubleshooting	49.4	Negotiation	6.9
Selective attention	11.5	Quality control analysis	46.0	Chemistry	5.7
Troubleshooting	9.2	Physics	44.8	Time Sharing	5.7
Technology design	9.2	Engineering and technology	42.5	Equipment selection	4.6
Programming	8.0	Operation and control	42.5	Social perceptiveness	4.6
High STEM/Low Soft occupations					
Repairing	22.6	Equipment maintenance	73.6	Oral expression	58.5
Equipment maintenance	15.1	Equipment selection	67.9	Negotiation	54.7
Spatial orientation	5.7	Repairing	66.0	Persuasion	52.8

Table 8.1 (continued)

"Very high" KSAs	Share of occupations (%)	"High" KSAs	Share of occupations (%)	"Low" or "very low" KSAs	Share of occupations (%)
High STEM/Low Soft occupations *(cont.)*					
Selective attention	3.8	Operation and control	56.6	Service orientation	49.1
Technology design	3.8	Troubleshooting	54.7	Social perceptiveness	49.1
Telecommuni-cations	3.8	Spatial orientation	50.9	Speech recognition	45.3
		Operation monitoring	45.3	Speaking	45.3
Mid STEM/High Soft occupations					
Medicine and dentistry	24.3	Monitoring	62.2	Operation and control	62.2
Management of financial resources	21.6	Active learning	59.5	Troubleshooting	50.0
Biology	20.3	Judgment and decision making	55.4	Equipment selection	43.2
Speech clarity	17.6	Speech recognition	50.0	Operation monitoring	31.1
Problem sensitivity	17.6	Instructing	48.6	Quality control analysis	29.7
Science	12.2	Written expression	48.6	Perceptual speed	17.6
		Reading comprehension	48.6	Chemistry	13.5
Mid STEM/Mid Soft occupations					
Programming	17.0	Computers and electronics	22.6	Operation and control	20.8
Telecommuni-cations	11.3	Management of material resources	20.8	Chemistry	20.8
Number facility	7.5	Management of personnel resources	18.9	Selective attention	17.0
Mathematics	7.5	Management of financial resources	17.0	Operations analysis	17.0
Mathematical reasoning	5.7	Medicine and dentistry	17.0	Equipment selection	15.1

(continued)

Table 8.1 (continued)

"Very high" KSAs	Share of occupations (%)	"High" KSAs	Share of occupations (%)	"Low" or "very low" KSAs	Share of occupations (%)
Mid STEM/Mid Soft occupations *(cont.)*					
Mathematics	5.7	Spatial orientation	17.0	Mathematics	11.3
Computers and electronics	5.7	Time management	17.0	Troubleshooting	11.3
Selective attention	5.7	Service orientation	15.1	Perceptual speed	9.4
Mid STEM/Low Soft occupations					
Spatial orientation	6.7	Equipment maintenance	56.2	Speaking	71.4
Repairing	3.8	Repairing	54.3	Oral expression	70.5
Equipment maintenance	1.9	Equipment selection	46.7	Persuasion	66.7
Selective attention	1.9	Spatial orientation	41.9	Speech recognition	62.9
Time Sharing	1.9	Operation and control	40.0	Social perceptiveness	62.9
		Operation monitoring	31.4	Reading comprehension	59.0
		Troubleshooting	30.5	Active learning	58.1
		Quality control analysis	20.0	Negotiation	58.1
Low STEM/High Soft occupations					
Speech clarity	34.0	Speech recognition	52.8	Troubleshooting	88.7
Social percep- tiveness	18.9	Writing	50.9	Operation and control	81.1
Speaking	13.2	Oral expression	50.9	Quality control analysis	77.4
Service orientation	11.3	Judgment and decision making	47.2	Equipment selection	77.4
Active listening	9.4	Monitoring	47.2	Operation monitoring	64.2
Persuasion	9.4	Reading comprehension	45.3	Engineering and technology	50.9
Learning strategies	9.4	Active listening	43.4	Chemistry	49.1
		Persuasion	43.4	Physics	47.2

Table 8.1 (continued)

"Very high" KSAs	Share of occupations (%)	"High" KSAs	Share of occupations (%)	"Low" or "very low" KSAs	Share of occupations (%)
Low STEM/High Soft occupations *(cont.)*					
		Written expression	43.4	Visualization	37.7
Low STEM/Mid Soft occupations					
Number facility	4.1	Speech recognition	24.5	Troubleshooting	58.2
Medicine and dentistry	4.1	Service orientation	21.4	Equipment selection	53.1
Speech recognition	3.1	Near vision	18.4	Operation and control	52.0
Speech clarity	3.1	Persuasion	13.3	Quality control analysis	50.0
Selective attention	3.1	Negotiation	12.2	Operation monitoring	48.0
		Number facility	10.2	Chemistry	48.0
		Speech clarity	8.2	Engineering and technology	46.9
Low STEM/Low Soft occupations					
Spatial orientation	3.5	Spatial orientation	13.5	Complex problem solving	79.4
		Equipment maintenance	9.2	Judgment and decision making	75.2
		Repairing	7.8	Active learning	73.8
		Speech recognition	6.4	Information ordering	71.6
		Time sharing	5.7	Fluency of ideas	69.5
		Operation and control	5.7	Systems evaluation	68.8
		Selective attention	5.0	Critical thinking	66.7
		Operation monitoring	5.0	Reading comprehension	66.0

least two standard deviations from the mean for all occupations) of six specific STEM skills. As can be seen, 26.3 percent of High STEM/High Soft occupations required a very high level of physics knowledge, and 24.2 percent required a very high level of biology knowledge. A little more than one-fifth (21.1 percent) of High STEM/High Soft occupations required a very high level of science skill. And one-fifth or more required a very high level of engineering and technology knowledge, design knowledge, and technology design skill. However, nearly one-fifth of High STEM/High Soft occupations (18.9 percent) also required a very high level of the Soft KSA problem sensitivity.

More than half of High STEM/High Soft occupations required a high level (at least one but less than two standard deviations above the norm for all occupations) of the Soft skills complex problem solving, judgment and decision making, and monitoring. Nearly half of High STEM/High Soft occupations required a high level of skills grouped into the STEM KSA bundle—systems analysis, science, and operations analysis—but a similar share required a high level of the Soft KSAs reading comprehension and information ordering.

Table 8.1 indicates that not all STEM KSAs are required at high levels, even in High STEM occupations. For example, several of the skills required at a low or very low level among High STEM/High Soft occupations were among the STEM bundle of KSAs. Roughly one of every eight High STEM/High Soft occupations required a low or very low level of equipment selection and operation and control skills.

Table 8.2 provides the largest levels and shares of employment for each STEM/Soft category accounted for by occupations requiring a high or very high level of the cross-cutting KSAs identified in Table 8.1.

Problem Solving and Judgment Characterize High STEM/High Soft Occupations

Table 8.1 provides the KSAs that are shared across the most occupations for each human capital category. As would be expected, occupations in the High STEM/High Soft category were most likely to require a very high level of knowledge related to specific scientific, engineering, and technology fields. However, the importance of being able to recognize and solve problems, as well as make judgments, was also clearly important. More than half of High STEM/High Soft employ-

Table 8.2 Employment and Share of Total Group Employment in Occupations Requiring Cross-Cutting High or Very High KSAs

Cross-cutting KSAs required at a "High" or "Very High" level	Employment	Share (%)
High STEM/High Soft occupations		
Operations analysis	7,397,260	75.8
Judgment and decision making	5,106,490	52.3
Monitoring	4,465,120	45.7
Information ordering	4,412,380	45.2
Systems analysis	4,246,150	43.5
Engineering and technology	4,118,580	42.2
Complex problem solving	4,049,900	41.5
Technology design	3,852,710	39.5
Science	3,782,350	38.8
Design	3,766,470	38.6
Physics	3,348,180	34.3
Reading comprehension	3,235,860	33.2
Problem sensitivity	2,773,170	28.4
Biology	1,768,470	18.1
High STEM/Mid Soft occupations		
Troubleshooting	5,682,120	65.1
Repairing	5,597,900	64.2
Equipment maintenance	5,427,120	62.2
Equipment selection	4,893,630	56.1
Quality control analysis	4,412,150	50.6
Engineering and technology	4,018,980	46.1
Physics	3,849,910	44.1
Operation and control	3,289,850	37.7
Selective attention	3,063,230	35.1
Telecommunications	3,045,650	34.9
Technology design	2,986,740	34.2
Programming	2,604,710	29.9
High STEM/Low Soft occupations		
Equipment maintenance	5,104,670	86.3
Spatial orientation	4,886,420	82.6
Repairing	4,868,860	82.3
Equipment selection	3,951,410	66.8
Operation and control	3,260,740	55.1
Troubleshooting	2,772,010	46.9
Operation monitoring	1,566,690	26.5

(continued)

Table 8.2 (continued)

Cross-cutting KSAs required at a "High" or "Very High" level	Employment	Share (%)
Mid STEM/High Soft occupations		
Speech recognition	9,502,680	63.7
Instructing	7,359,480	49.3
Monitoring	7,317,830	49.0
Medicine and dentistry	5,370,710	36.0
Problem sensitivity	5,074,960	34.0
Judgment and decision making	4,875,860	32.7
Management of financial resources	4,513,110	30.3
Biology	4,457,230	29.9
Written expression	3,899,770	26.1
Speech clarity	3,785,560	25.4
Active learning	3,207,580	21.5
Reading comprehension	2,878,090	19.3
Mid STEM/Mid Soft occupations		
Management of financial resources	3,344,960	42.6
Management of material resources	3,299,810	42.0
Management of personnel resources	2,902,000	37.0
Time management	2,896,250	36.9
Service orientation	2,567,060	32.7
Medicine and dentistry	1,614,730	20.6
Computers and electronics	1,581,280	20.1
Programming	1,485,810	18.9
Telecommunications	1,351,660	17.2
Spatial orientation	998,590	12.7
Mid STEM/Low Soft occupations		
Repairing	4,768,690	71.9
Equipment maintenance	4,338,740	65.4
Spatial orientation	4,154,130	62.6
Operation and control	3,599,290	54.3
Equipment selection	1,933,870	29.2
Troubleshooting	1,439,870	21.7
Operation monitoring	1,323,240	20.0
Low STEM/High Soft occupations		
Speech recognition	3,770,720	71.43
Persuasion	3,650,640	69.16
Judgment and decision making	3,479,120	65.91
Active listening	3,102,610	58.78

Table 8.2 (continued)

Cross-cutting KSAs required at a "High" or "Very High" level	Employment	Share (%)
Low STEM/High Soft occupations *(cont.)*		
Social perceptiveness	3,039,940	57.59
Service orientation	3,021,900	57.25
Writing	2,892,240	54.79
Oral expression	2,507,610	47.50
Speech clarity	2,086,470	39.53
Speaking	2,062,950	39.08
Monitoring	1,999,800	37.88
Learning strategies	1,854,270	35.13
Reading comprehension	1,812,410	34.33
Written expression	1,671,480	31.66
Low STEM/Mid Soft occupations		
Persuasion	11,287,550	40.50
Negotiation	10,919,940	39.18
Service orientation	10,864,080	38.98
Speech recognition	6,805,410	24.42
Number facility	5,086,460	18.25
Low STEM/Low Soft occupations		
Spatial orientation	6,384,930	14.18
Time sharing	2,483,360	5.51
Selective attention	1,475,620	3.28
Equipment maintenance	1,318,100	2.93

SOURCE: O*NET and OES (2014).

ment required a high or very high level of judgment and decision making; nearly half of employment required a similarly high proficiency in assessing performance (monitoring, 45.7 percent) and observing patterns (information ordering, 45.2 percent).

The employment distributions in Table 8.2 reinforce the high cognitive demand of High STEM/High Soft occupations. Although roughly 40 percent of employment in the category was in occupations requiring high or very high levels of engineering and technology knowledge and science skill—two KSAs typically associated with STEM jobs—three-fourths of all employment required a high or very high level of a less-celebrated STEM skill, operations analysis. O*NET defines operations analysis in terms of the technical skills needed to create designs. This is a skill shared by engineers, physicists, and computer systems analysts, as well as business operations specialists, natural sciences managers, and architects.

Equipment Expertise Characterizes Other Occupations Requiring High STEM

As can be seen in Table 8.1 and might be expected given the results of the regression analyses, there is a degree of skill overlap among the High STEM/Mid Soft and High STEM/Low Soft occupations. Most occupations in both categories are involved with selecting, maintaining, and repairing equipment, as well as troubleshooting potential problems. Nearly 90 percent of High STEM/Low Soft occupations—and more than 80 percent of employment—require a high or very high level of repairing and equipment maintenance expertise.

What seems to separate the two occupation categories is knowledge of physics, engineering and technology, and programming skills. More than 42 percent of High STEM/Mid Soft occupations require a high level of engineering and technology expertise, and 8 percent of such occupations require a very high level of programming skills. Nearly 30 percent of High STEM/Mid Soft employment requires high or very high programming skill (Table 8.2). Another distinguishing characteristic is the low or very low level of communication skills associated with High STEM/Low Soft occupations.

Mid STEM Categories: A Mashup of Medical, Computer, and Numeracy Capabilities

A very high level of medical and management expertise characterizes Mid STEM/High Soft occupations. Such occupations tend to require a high capacity for assessing, monitoring, and conveying information. Mid STEM/Mid Soft occupations are characterized by a high or very high level of computer knowledge, programming skills, and numeracy. Mid STEM/Low Soft occupations seem somewhat similar to High STEM/Low Soft occupations in their demand for a relatively high level of repairing and equipment maintenance abilities and a low level of communication skills. Occupations with such skill demands account for two-thirds or more of all Mid STEM/Low Soft employment (Table 8.2). This suggests that regions with demand for High STEM/Low Soft workers may find a pool of talent to tap among Mid STEM/Low Soft occupations.

Communication and Judgment Central to Low STEM/High Soft Occupations

A high or very high level of communication skills characterizes Low STEM/High Soft occupations. Such occupations require workers to be able to listen, learn, and make decisions. Moreover, these are occupations that require workers to speak and act persuasively. Two-thirds or more of all Low STEM/High Soft employment requires a high or very high level of speech recognition, persuasion, and judgment and decision making (Table 8.2). More than half of Low STEM/High Soft employment is in occupations requiring a high writing level. Low STEM/High Soft and Low STEM/Mid Soft occupations both can be characterized, in part, by their lack of engineering knowledge and troubleshooting skill. The high or very high level of number facility in certain Low STEM/Mid Soft occupations may provide a foundation to move workers into Mid STEM/Mid Soft occupations.

Low STEM/Low Soft Occupations Have Low Expectations for Thinking and Deciding

As would be expected, Low STEM/Low Soft occupations are largely characterized by what they lack—two-thirds or more of such occupations require a low or very low level of problem-solving, critical thinking, and decision-making skills. In other words, what seems to set these occupations apart is the low expectations for higher order thinking. Despite these occupations falling into the bottom category for skill demands, nearly one in five requires high or very high spatial orientation capabilities. This translates into some 6.4 million workers, ranging from bus drivers, delivery drivers, and industrial truck operators to hoist and winch operators, packaging machine operators, and postal service mail carriers.

CONCLUSION

Certainly, Tables 8.1 and 8.2 indicate that wide skill variation and broad overlap exist within and among the categories. This would seem to capture the nature of career pathways—not necessarily rigid and upright, as with the old "career ladder" image, but branching and potentially circuitous. However, broad themes about human capital concentrations and deployment do emerge. These demand-based themes presumably offer more concrete direction for efforts to develop and expand human capital—for individuals, educators, employers, and policymakers alike—than does the current, narrow focus on STEM degrees.

Without a doubt, workers in diverse fields and industries develop specific expertise and proficiencies unique to the products, processes, and activities of those fields and industries. This analysis does not suggest that the work of aircraft mechanics and service technicians and the work of computer network architects is similar because both occupations require a high level of repairing skills. But it does aim to broaden understanding of and appreciation for "high" skill beyond the perception evident in public policy and the popular press. The analysis also attempts to highlight the connective tissue between occupations that serves as a foundation for professional advancement, an avenue for career change, and a safety net for job disruption.

Looking across all three High STEM categories, it is interesting to note that several occupations in all three groups require a very high level of technology design capability. This suggests that workers in such High STEM/Low Soft occupations may find they can access somewhat better-paying High STEM/Mid Soft jobs by improving communication skills. Workers in High STEM/Mid Soft occupations may find that they can improve access to better-paying jobs by working to enhance—through education or experience—their engineering and technology, programming, and technical design capabilities, as well as their problem-solving and decision-making skills. Regions that seem to experience a shortage of workers with skills in specific occupations may find they can build on cross-cutting skills to more efficiently retrain and redeploy human capital.

The high or very high level of number facility required of certain Low STEM/Mid Soft occupations may provide a foundation to move workers into potentially better-paying Mid STEM/Mid Soft occupations. Alternatively, some of the 11.3 million Low STEM/Mid Soft workers in occupations requiring a high level of persuasion skills may find that developing that human capital asset allows them to transition into higher wage Low STEM/High Soft jobs.

Several Low STEM/Low Soft occupations require a relatively high level of equipment maintenance and repairing skill. Efforts at helping low-wage workers to access career paths may be most successful if they build on technical skills or they focus on developing problem-solving and communication capabilities that are largely absent from the expectations of Low STEM/Low Soft occupations.

Beyond the insight it potentially affords for career pathways, this deconstructing of human capital categories into their characteristic high components makes a case for greater policy focus on developing critical soft skills. Business literature and anecdotes support the value of such skills, and the high-order thinking skills associated with engineering, scientific, and other High STEM/High Soft occupations may help to explain why businesses are increasingly hiring STEM majors for non-STEM occupations, particularly management and professional ones (Carnevale, Smith, and Melton 2011). Yet, making decisions, solving problems, observing patterns, assessing performance, and reading deeply are not simply by-products of STEM majors, nor are they exclusive to them. In fact, these skills may be as critical to technological innovations, new products, and new businesses as the more technical expertise so many STEM policies seek to promote.

9

Concluding Thoughts and Policy Implications

In early 2017, Kentucky legislators hammered out a performance-based plan for funding postsecondary education, joining nearly two-thirds of states in tying public support to a desired set of outcomes. The Kentucky plan, which will be phased in through 2022, adheres to Gov. Matt Bevin's broad framework for state investment in human capital development: more electrical engineers and fewer French literature majors. The plan incentivizes universities to produce more graduates with degrees in science, technology, engineering, math, and health (Papka 2017). The bill also rewards two-year institutions for producing more workers with credentials linked to industry demand.

Pursuit of STEM degrees has moved from one of personal interest or professional ambition to a matter of economic imperative and public priority. The policy assumption is clear: economies benefit from more scientists making discoveries, more engineers solving problems, and more computer experts programming solutions.

Yet, many STEM initiatives adopted by state and local governments have little regard for the differences of place. Largely imitative policies that focus on increasing the supply of technical talent frequently assume a relatively uniform demand for such talent across states and metropolitan regions. The intense interest in STEM degrees obscures the importance of other types of knowledge and skill, particularly in the context of place, and overshadows other paths to developing in-demand capabilities. At a time when technology has broadened the reach of firms and individuals to engage in world markets, ideas and information are exchanged virtually instantaneously, and change happens at a rapid pace, the path to developing the knowledge and skills needed to operate within this environment has narrowed to the pursuit of one of a handful of educational degrees that take years to achieve.

This research provides support for a multifaceted view of human capital based on how educational attainment, technical knowledge, and interpersonal skills are demanded and valued in the marketplace. The analysis suggests that an alternative measure of human capital reflecting the skill sets required of a region's collection of occupations may offer greater insight to policymakers and practitioners tasked with supporting and improving regional economic performance than the common focus on educational attainment of the area's population. This is especially true if policymakers and practitioners are interested in measures of economic well-being other than regional median wage.

This analysis demonstrates that having a larger share of residents with a bachelor's degree or higher does correspond with a higher regional median wage, as human capital theory advances. However, such high levels of advanced educational attainment is not necessarily closely associated with other important measures of regional economic performance, such as GRP growth, increased total factor productivity, and reduced poverty rates. These results seem to be somewhat in conflict with the largely rosy conclusions from a simplified rendition of human capital theory, but they bolster the frequently ambiguous and often problematic empirical findings in the literature and in practice. Equivocal findings suggest that either human capital theory is more nuanced than assumed or the variable commonly used to measure it is not up to the task—or both.

Matching the extensive details on occupational requirements now available through the government-sponsored O*NET to occupational and region-specific data collected by the federal government through the Department of Labor and the Census Bureau provides the means to explore whether a finer grained measure of regional human capital acquisition and deployment has the potential to reveal the pathways to theorized economic benefit. Such a measure allows for a more nuanced understanding of occupations as a bundle of attributes.

Measuring human capital as the collection of knowledge, skills, and abilities (KSAs) required of occupations has two important advantages over the commonly adopted proxy for human capital, educational attainment. First, conceptualizing human capital through KSAs more closely captures the broad concept of human capital as conceived by Schultz (1961) and as observed in economic literature as far back as Adam Smith (1776/2008). Smith, Schultz, and Becker (1964/1993) all depict human capital as superior knowledge and skill, no matter how acquired. Second, it squarely acknowledges that human capital is a factor of production, meaning that the value of human capital extends from how it connects to the economy. This in no way minimizes the broader value of education, which has been shown to be associated with a number of desirable outcomes ranging from healthier living to increased voting rates. However, in the practical world of public policy, limited resources presumably should be applied to best effect. Human capital investments that are over-allocated toward education—or specific types of education—rather than better matched to the human capital demands of a region mean that other potential human capital investments, such as developing "soft" interpersonal skills, improving the health of families, or maintaining a safe environment, may go underfunded.

Certainly, elevating the potential of its people is an important role for government. However, human capital theory assumes that such investments yield economic return, bringing benefit to those who pursue "superior skills," whether they are individuals investing private resources or governments investing public ones. This, by extension, means the human capital investments are in some way creating greater economic value. Resource-based theory of the firm, which has roots in the economic literature but has been explored more extensively in the business literature, provides an important framework for regional (and state) policymakers regarding how the regional human capital asset contributes to value creation and sustained competitive advantage.

However, understanding opportunities for value creation and sustained competitive advantage requires better understanding of a region's human capital assets. In other words, it requires awareness of opportunities that may arise out of differences in regional human capital deployment, instead of the largely imitative policies that assume relatively similar approaches to regional human capital development will yield desired results.

KEY TAKEAWAYS

Four observations of import for workforce and education policy leap from the pages of this book: 1) What is often missing from regional economic development policies and programs is an understanding and acknowledgment of how specific skills are affected by the rise and fall of the industries that demand them. 2) Focusing policy attention keenly and narrowly on a rather imprecise perception of what constitutes in-demand skills, without understanding which components of an occupation are actually demanded by employers, may lead to distortions in the supply of skills as education and training systems respond to a false set of signals. 3) Mischaracterizing the skill makeup of occupations and jobs contributes to misperceptions about career ladders and emphasizes educational attainment over skill achievement. 4) Too often, the "blame" for low levels of human capital development is placed on workers without conceding how many occupations actually demand very little skill.

This book began with a series of questions reflecting the complexities and contradictions evident when national and state goals for increasing the supply of STEM degrees are adopted and applied locally: Does a larger share of STEM-degreed workers really improve the regional economy? Do regions have similar demand for such talent? Does promotion of STEM degrees—or degrees in general—neglect other avenues for workforce investment? What are the human capital best bets that can be made to address regional workforce challenges, align with opportunity, and advance regional economic well-being?

Answers to these questions and other key findings are summarized below.

A region's human capital assets come from how the knowledge, skills, and abilities of individual workers are developed and then deployed through the region's mix of jobs.

The resource-based literature suggests that a region's economic well-being arises out of how valuable, rare, inimitable, and apropos its human capital assets are within the context of its mix of industries. Much of the discussion of regional human capital in the economic development literature focuses on some measure of educational attainment of individuals. However, a region's individual-level human capital capacity also includes worker skills developed through training, practice, or self-study, as well as worker experience and even health. A region's human capital asset base, its portfolio of economically valued skills and knowledge, is continually adjusted through the ebbs and flows of migration and the attributes of new entrants into the labor force. A region's human capital asset base also encompasses firm-level human capital, which includes firm-specific practices and processes, work-related intellectual property, and organizational systems and structures. Both individual- and firm-level human capital have value in their own right, and each is a building block of a region's portfolio of human capital assets.

Regional human capital assets, whether measured by educational attainment (degree) or by occupational skill requirements, vary widely across regions.

The average level of college completion across metropolitan regions was 26 percent, but that average belies considerable variation among them. More than 45 percentage points separate the region with the lowest share of college degree attainment from the region with the highest. Yet, this wide range in regional degree attainment may, at least in part, reflect wide variation in each region's mix of occupations. There was a five-fold difference in the share of regional employment in occupations requiring both above-average STEM and above-average Soft skills, with the least highly skilled region employing one of every 20 workers in such occupations and the highest skilled region employing one of every four. Conversely, some regions had more than 60 percent of their workers employed in occupations requiring below-average STEM and below-average Soft skills, whereas other regions had little more than 30 percent of workers in such low-skill jobs.

Measuring regional human capital in terms of occupational skill requirements explains differences in regional well-being better than the use of college degree attainment.

The regression analyses substituting the occupation-based human capital variables consistently explained variation in the five measures of regional economic well-being better than the population-based educational attainment variable. Even the relatively blunt grouping of occupations by above-average or below-average STEM and Soft skill requirements substantially improved explanatory power over the educational variable based on degrees. This would seem to be expected, given that occupations are the means by which human capital is connected to the economy. Refining the occupational variables continued to improve the explanatory power of most of the regression models of five measures of economic well-being examined in this study. For example, although the education variable was statistically significant in predicting median wage, the model in which it was added to four control variables explained only about 60 percent of regional variation in median wage, compared to the 82 percent of variation explained by the control variables and the nine occupation-based variables.

Increasing the share of a region's population with a bachelor's degree or higher may improve some measures of regional economic performance but may not affect, or may even worsen, others.

Consistent with human capital theory, regions with a larger share of highly educated adults tend to have higher median wages than less-educated regions. However, regions with higher levels of education did not enjoy greater GRP growth rates, higher productivity, or higher per capita incomes than less-educated regions, after controlling for labor force participation, cost of living, manufacturing employment, and population growth rates. Moreover, a "mismatch" between the share of the population with a bachelor's degree or higher and the share of a region's occupations requiring such level of educational attainment may slightly lower regional wages, while slightly increasing growth in GRP. This last finding would be consistent with a regional labor market that has an abundant supply of potentially skilled workers, where lower wages attract demand leading to higher than expected GRP growth rates as employers expand to take advantage of a relative bargain in the labor market or new employers are attracted into these markets.

Increasing the share of a region's population with a bachelor's degree or higher in STEM does not necessarily improve regional economic well-being.

When policymakers and reporters tout the importance of STEM skills and STEM jobs as drivers of innovation and economic growth, they typically are referring to occupations that require both higher than average STEM capabilities and higher than average thinking and communication skills. Nearly three-fourths of the 182 occupations grouped in this category require a bachelor's degree or higher. However, High STEM/High Soft occupations account for little more than 16 percent of total U.S. employment. Refining the human capital measure further to include only those occupations requiring STEM and Soft skills in the top third of occupational skill demands for each KSA category reveals that 83 percent of such occupations, accounting for only about 4.3 percent of total U.S. employment, require a bachelor's degree or higher. Occupations such as physicists, computer network analysts, microbiologists, and engineers of all stripes fall into this category of High STEM/ High Soft requirements, as do information security analysts, chemistry professors, and nurse practitioners. Regions with a higher than average share of employment in High STEM/High Soft occupations enjoyed higher regional median wages and higher productivity, but such concentrations were shown to have no effect on per capita income or poverty rates.

Occupations requiring higher than average STEM skills are important to regional economic performance, but such occupations may not require a college degree.

Although national, state, and regional policies targeted toward increasing the supply of workers with STEM knowledge have displayed a bias toward encouraging increased higher educational attainment (Rothwell 2013), regions with a larger share of employment in occupations requiring STEM KSAs in the top third of all occupations but Soft skills in the bottom third saw gains across all five measures of economic well-being during the study period. However, such occupations account for only about 5 percent of total U.S. employment, and they include many traditional "blue-collar" occupations, such as derrick operators and roustabouts for the oil and gas industry, industrial machinery mechanics, and machinists, none of which requires a bachelor's degree. STEM initiatives directed at occupations requiring skills beyond that of a high school diploma but less than a four-year college degree have been increasing, against a backdrop of anecdotal reports from manufacturers and advocacy groups indicating a need for workers with such skill sets. These findings support those efforts, provided that they are aligned to regional demand.

The focus of human-capital based policy interventions is typically on increasing the supply of highly skilled workers, but the share of regional employment in occupations with the lowest skill requirements represents a stubborn challenge to economic well-being.

Occupations with requirements in the bottom third of STEM and Soft skills account for 18.4 percent of all occupations, but they account for 34.1 percent of U.S. employment. Such employment is associated with lower individual wages, lower regional median wages, lower GRP growth, lower total factor productivity, and lower per capita incomes. Policies that focus on increasing the supply of highly skilled workers will do little to alter these occupational low-skill expectations and the associated impacts on workers and regions.

Human capital accumulation that most benefits regions may not be that which most benefits individual workers.

Although regions with a higher share of employment in High STEM/Low Soft occupations saw improvements on all measures of regional economic well-being, such occupations paid a median wage of only $39,100, which ranked near the bottom of the wage scale for the nine human capital STEM/Soft categories. The highest STEM/Soft category paid individuals the highest median wages by far—$79,930— even though their benefit to regions was less pronounced and broad based. Occupations requiring Low STEM/High Soft skills paid median wages of $61,450; however, regions with a higher share of employment in such occupations saw increases in regional median wage but no improvement in the other measures of economic well-being, as demonstrated in Chapter 7.

The regional human capital asset base is important, but it only explains part of the reason why some regions perform better than others.

Although the occupation-based human capital measures improved the explanatory power of all of the models, there was still substantial variation in the measures of regional economic well-being left unexplained. Roughly one-third of the variation in per capita income, nearly half of the variation in poverty rates, and two-thirds of the variation in regional GRP growth rates could not be explained by regional differences in labor force participation rates, cost of living, manufacturing employment, population change, and occupational human capital. This unexplained variation suggests that, at least in the short run, a combination of business cycle movements, firm dynamics, and industrial legacies exerts a large influence on the economic well-being and the performance of regional economies.

LIMITATIONS

Although the findings presented here offer a more refined and robust understanding of regional human capital, they should be viewed somewhat cautiously. The results are a snapshot in time. The way O*NET and OES data are collected inhibits the comparison of regional skill sets and economic performance over time. Moreover, the need to use five-year ACS data to match to the MSA delineations used by OES means that the measures of economic well-being were still being affected by the long-lingering effects of the Great Recession, which officially ended in summer 2009. It is reasonable to assume that such a far-reaching and deep economic disruption has lasting results. For example, the occupational skill categories associated with improved economic well-being may simply reflect high concentrations of industries that experienced quicker or more pronounced bounce-back from the effects of the recession, such as oil and gas production, automobile manufacturing and its supply chain, and the financial services industry.

Assumptions regarding the uniformity of occupational skill sets across industries and regions may represent serious limitations of this research. O*NET's use of only a couple dozen workers to represent the human capital requirements of occupations across the nation assumes a homogeneity of occupational human capital demand. Moreover, this analysis explores skill out of the specific context of place. Presumably, different areas may have different demands and pay different rewards to human capital. Workers with unique skill sets may not see the national average return on the investment in acquiring human capital if they live in an area where there is no demand for such skill. A better understanding of region-specific variation in occupational skill demands than is currently available in the O*NET database would improve on the findings presented here.

Despite these limitations, this research offers a warranted reframing of human capital study and practice away from an overriding focus on educational supply toward knowledge and skill demand. By offering a methodology specifically aligned to the current policy focus on STEM, this research expands the understanding of STEM jobs and their impact on regional economic well-being. Findings demonstrate that many traditional "blue collar" occupations require a relatively high level of STEM expertise and have become jobs where heads and hands are joined in producing value. Although frequently overlooked in STEM discussions and policies, such occupations are of critical importance to the overall health of regional economies.

The research also reveals the breadth of demand for Soft skills. Problem solving, communication, and decision making are among those skills that are required of a wide swath of occupations and that tend to reward workers who possess them. This research makes a case for greater policy focus on shared skills, whether technical or soft, that cut across occupations and industries as avenues for job access and career advancement.

Yet, better mapping of career pathways does little to address the prevalence of low-skill work in the economy. The results presented here highlight the limitations of human-capital-based interventions to offset the broadly negative effects on regional economic well-being of large concentrations of low-skill employment. Given that other studies have shown job growth to be occurring primarily at the low end of the skills and wage spectrum, policymakers may need to look for other ways to lift up their workers than through raising their skills.

POLICY IMPLICATIONS

People work in occupations that are directly linked to the provision of goods and services and the industries that provide them. In other words, occupational employment is derived from products and services that are produced, be it by a business or a unit of government. The connection between employment and output has important implications for policy interventions targeted at increasing the supply of human capital. Regions (or states and even nations) that invest in developing human capital that does not fit the demand for skills that is generated by the region's industrial mix will likely not reap the full benefit from public expenditures that underwrite its formation. Workers with ill-fitting human capital will either accept jobs that are typically staffed by someone with lower educational levels than they have acquired, or they will relocate to regions where the skills they possess match those that are in demand. Either scenario means the investing area will see little return on its human capital investment.

Economic development and workforce policy and practice have taken a largely supply-side view of human capital, assuming that increasing the educational levels of the population, especially increasing the share of workers with expertise in STEM, will be rewarded with economic benefits—whether measured as growth in GRP, improved incomes, or lower poverty rates. Such policies and practices are shaped by both observation and theory regarding the economic primacy of knowledge and technology. However, this view neglects the importance of demand, goodness of fit, and strategic deployment in transforming a region's supply of human capital into a valuable asset for regional economic well-being.

In the business literature, the resource-based theory of the firm places human capital as central to value creation and sustained competitive advantage. However, how those assets are developed and deployed is determined within the context of firm strategies, strengths, and capacities to respond to external market forces and seize on opportunities. Competitive advantage is achieved not simply through differences in resources; it is based on how those resources are used and the margin that employees generate for their employers. This suggests that human-capital-inspired economic development policies will not achieve the desired boost in economic well-being unless they are aligned with the occupational demand generated by a region's employers. Interventions that focus on regional human capital production, instead of regional human capital deployment, are likely to lead to distortions in the equilibrium between occupational supply and demand and miss opportunities to facilitate fit.

Further complicating policy efforts is an apparent conflict between the economic return for a region—or state or nation—overall and the possible return for individuals. For example, regions with a higher share of employment in High STEM/Low Soft occupations saw benefit across all five measures of regional economic well-being, at least for the period studied, but workers in such occupations earned relatively low wages. What is clear is the importance of superior Soft skills to worker wages: occupations falling in the top third in terms of Soft skill requirements pay substantially more than all other occupational skill categories. Yet, larger concentrations of such skills had limited influence on the measures of regional economic well-being.

Regions that are fortunate enough to be home to industries that are growing instead of declining will see greater economic benefit as their supply of human capital better matches industrial demand. Instead of adopting broad "me-too" policies targeted toward producing more bachelor's degrees, specifically STEM degrees, regions would be wise to focus economic and workforce development efforts on human capital "fit." Good human capital fit allows regions to seize the gains that accompany industries that are experiencing periods of growth. That means supporting occupations and specific skills that support regional industries as long as those industries and their products are growing or gaining market share.

However, fitting the existing occupational structure is a less desirable approach if the region's industrial base is in decline or transition. The problems of "Rust Belt" cities, where skill sets were too closely aligned to a handful of dominant industries, demonstrates this point. A second lesson can be drawn from the Rust Belt experience: in many of those regions, manufacturing is competitive, but its occupational skill needs and employment levels have changed, in part owing to automation.

Human-capital-based interventions more aligned to the specific needs of industry invite questions about the appropriate role for government. In his well-known essay on education, Milton Friedman (1955) suggested that public support should be more directed at the types of broad knowledge that contribute to citizenship and leadership and cautioned against public support for varieties of human capital where the benefits are mostly captured by the individuals (and, presumably, firms). For example, firm-specific tacit knowledge and skills are of the greatest value to firms and are best taught in the workplace.

This suggests that regions (and states and nations) should focus on the fungibility of their human capital stock, which may provide regions with the ability to adapt when the product cycle of an industry begins to plateau or decline. Friedman's reasoning fails to recognize the potential value of knowledge spillovers, which may result in societal benefit from investments in human capital beyond the observed private benefit. However, his essay offers important insight into the delicate balance policymakers face. In the practical world of policy, limited public resources presumably should be applied toward those efforts that promise highest public benefit.

Adding to this delicate policy balance is the need to be aspirational while also practical, the need to anticipate the human capital requirements of tomorrow while supporting the immediate demands of today. This is indeed a challenging balance to strike, especially in an environment of rapid technological change, intense global pressures, and political expectations of action. What seems clear, however, is that countless human-capital-based economic development initiatives, especially at the regional (and state) level, are being undertaken with an incomplete or misguided understanding of how such efforts help to grow a regional human capital asset of greater economic value.

This research suggests a number of strategic and tactical adjustments in human-capital-focused policies and programs that are targeted toward improving regional economic well-being. Although the focus of this research is on the regional economy and the audience for the recommendations below consists of local policymakers, practitioners, and entities that provide education and training, having a regionally responsive workforce system relies on more than local actors alone. A considerable challenge to effective regional economic development policies is that regional economies often extend beyond local political jurisdictions, and the funding for local workforce efforts often reflects state and federal political priorities. What follows are eight rules, based on this research, that should improve the effectiveness of regional workforce efforts.

1) Human-capital-based policies and interventions implemented to improve the regional economy should be data driven.

Investments in a region's human capital assets are more likely to yield better economic returns if they are based on region-specific information regarding current employment levels, as well as short-term and long-term trends. A regional database of occupational requirements and industry needs would be valuable in helping to shape decisions about which types of human-capital-based investments of public resources are most appropriate and represent the best return to the region. In an analysis of large, publicly traded businesses, Brynjolfsson, Hitt, and Kim (2011) found significant improvement in output and productivity among firms that used data to drive decision making. Presumably, regions relying on data to drive decisions about economic development and workforce development policies and interventions would see similar benefit.

Certainly, creating a region-specific inventory of occupational and industry human capital assets and needs may be beyond the resources of many regions to construct. However, there is a wealth of information about current levels of employment, as well as short-term and long-term projections, available through federal and state databases. Occupational employment can be matched through the Standard Occupational Classification (SOC) system, and data on industry employment can be matched using the North American Industry Classification System (NAICS). It is challenging to match employment data available through the two coding systems, but the National Employment Matrix incorporates data from the OES, Current Employment Statistics, and the Current Population Survey and aligns, or crosswalks, the occupational measures to industry employment data. This allows trends in occupational employment within specific industries to be examined. As the name suggests, and similar to the challenges discussed regarding O*NET, the National Employment Matrix is a nationwide database that lacks the level of detail to explore specific regional occupational differences. However, the matrix does provide insight into which occupations actually exist within certain industries. Drawing on the information available through O*NET and the National Employment Matrix can provide policymakers and practitioners with a roadmap for developing an integrated database of how human capital is deployed throughout their region's mix of occupations and industries. It will also provide information on the specific composition of KSAs within the region's industries.

2) Focus on public policy "small ball."

Baseball fans are thrilled by teams with home-run power. The seemingly boundless hope of winning a game with a single swing of the bat captures the imagination and brings an exciting rush. Yet, swinging for the fences often results in strikeouts—lots of them. Teams that employ "small ball" tactics—getting on base, advancing the runner, making plays by being quick, nimble, and observant—may lack the captivating allure of high-scoring slugfests, but they win games. Such sound fundamentals are especially important in tightly competitive contests and when players lack the natural advantages of power hitters.

Think of President Obama's challenging the nation to increase the number of STEM workers as a long-ball play. Certainly, it is reasonable to assume that the rapid technological change and intensely competitive environment of the past several decades will continue. It is also reasonable to expect that the nation will need a continuing supply of well-trained workers with the skills to match the constantly evolving demands of employers and to continue the technological and product discoveries that help to drive economic growth. Expanding the nation's roster of power hitters increases the likelihood of economic home runs (this is where the law of large numbers meets regional economic development planning). The nation as a whole has a greater capacity to absorb the inevitable strikeouts along the way. It also has an easier time retaining the top talent it has invested in developing because national borders are semi-binding constraints.

Yet, regions—and, to a certain extent, states—do not have the same access to reserves of talent. They also have fewer options for allowing top talent, when developed, to operate at its highest level and have even less ability to retain heavy hitters who have better prospects elsewhere. This in no way minimizes the important role of regions in supporting the development of skills and abilities that ultimately help to build a competitive national economy. However, it does suggest a different approach to talent development and acquisition. When regions adopt initiatives that largely imitate the successes of other regions without the underlying labor demand or industry mix to support them—think about the many efforts to grow the next Silicon Valley—they engage in the policy equivalent of "swinging for the fences." The approach may occasionally yield results, but more frequently it will not. Instead, regions need a more deliberate, methodical approach to developing and deploying their human capital resources that emphasizes sound fundamentals, seizes on incremental opportunities for advancement, and adapts to changing matchups on a competitive field of play.

Taking this approach to human capital from analogy to practice requires the kind of evidence-based understanding of the regional economy advocated in the first recommendation to serve as a foundation for investment. Data-driven analysis should help regions "manufacture runs" by supporting smart investments in emerging employment opportunities and defending against anticipated skill gaps, such as the predicted shortage in manufacturing and trades workers due to an aging workforce nearing retirement.

3) Resist the urge to imitate, act on perceptions, and rely too heavily on the accounts of the business community or industry advocates when crafting human-capital-based policies.

Certainly, there is value in copying best practices that have been demonstrated to be effective in other regions; perceptions may reflect reality, and the vantage point of business offers important insight about near-term labor demand. The regional human capital asset portfolio, after all, does not simply rest with individual workers. Rather, it reflects the interconnection between workers and firms. Even so, employers and trade associations are often quick to try to off-load their responsibilities for human capital development onto individual workers and public institutions. Employers must recognize, and accept, the important role they play in developing the human capital that they need to compete. As noted earlier, tacit, firm-specific skill is likely to be the most valuable to businesses and, thus, it is appropriately provided by businesses themselves.

Moreover, perceptions that distort reality frequently result in less-than-desirable individual choices and public policies. For example, Teitelbaum (2014) pointed to six decades of boom-and-bust cycles fueled by recurrent perpetuations of a "STEM myth." This myth foments a tendency among policymakers to equate growth in occupations requiring computer and technical skills with an increase in the general level of skill demanded by employers. Certainly, computerization may require a higher skill set to perform some occupations, but computer systems make many jobs easier to perform. For example, optimization programs reduce or eliminate the decision making of dispatch workers; self-ordering technology minimizes much of the communication skills required of wait staff; and computer diagnostic systems lessen the problem-solving skills and technical expertise required of vehicular mechanics. Workers need training to learn specific systems and tools, but many need only a certain familiarity with computer technology to perform their jobs. Computer skills are so ubiquitous now, especially among the newest generation of workers, that even those who lack a computer in their homes may have smart phones and be comfortable using technology.

Misperceptions about the demand for talent in the local—and even national—economy lead young people, families, and governments to overinvest in education for which there is insufficient demand. Policies that are in large part cheerleading and aspirational, and not directly linked to corresponding work opportunities and financial rewards for undertaking additional education and training risk doing real harm to individuals and communities alike. The costs in money, time, and forgone opportunities preclude the ability to make other investments that could be more efficient and effective in improving personal or community well-being.

4) Develop programs and support opportunities for internships, mentoring, and apprenticeships. Opportunities for hands-on learning and career exposure should particularly be directed at the high school level.

In 2013, the Ohio higher education system implemented a program incentivizing public four-year and two-year colleges to engage with local employers in industries identified as central to the state or regional economies in an effort to place undergraduate students in internships. The program was particularly targeted toward small businesses that had little to no experience with internships. Drawing on a portion of new casino licensing fees allocated to workforce development, $11 million was awarded to universities, and their nonprofit partners, through competitive processes. To encourage participation among employers, funds could be used to rebate a portion of the student intern's wage. Funding for faculty monitoring of the internship experience was also available. Given that internships are often a pathway toward job placement after graduation, this is an example of how human-capital-based initiatives can be tailored to improve the odds that the investments made by state and regional taxpayers stay in the region. Employers use internships as a way of trying out new talent, and students who are exposed to opportunities to apply their freshly honed talents within the region are more likely to stay.

Similar programs that offer experiential learning opportunities and structured career access for high school students are especially warranted, given the precipitous decline in teenage connection to the job market. According to a report from the Brookings Institution, employment rates among 16- to 19-year-olds fell from 35 percent in 2008 to only 29 percent in 2014 (Ross and Svajlenka 2016). This trend is troubling because disconnected teen workers often become disconnected adult workers. Moreover, connecting to the workplace as teenagers should help young people develop the job-readiness and other soft skills that employers frequently say they lack, as well as provide them with insight that can help to shape their decisions about educational endeavors and career paths.

Some employers, especially manufacturers, have recognized the need to reach out to high school students. For example, Conexus Indiana, an initiative focused on the workforce needs of the advanced manufacturing and logistics industries, launched a six-week summer internship program a few years ago for teenagers who participated in high school engineering and manufacturing programs. Students were paid while getting hands-on experience working with participating employers.

In 2014, the White House announced a nearly $500 million competition to encourage partnerships between local businesses and community colleges to develop apprenticeship programs. Such structured programs that combine classroom instruction, hands-on learning, and on-the-job mentoring are proven, effective means for allowing people to earn as they learn the skills they need to succeed. After more than a decade of precipitous decline in industry-connected workforce investment, it is heartening to see policy focus on increasing pathways to skill development. However, the nationwide total of 448,000 apprentices by the end of 2015 was minuscule compared to the more than 10 million students enrolled at four-year colleges and universities. Resistance to apprenticeships is, at least in part, one of perception. Apprenticeships are seen as largely paths to "blue collar" careers (Weber 2014) even though the Department of Labor's Registered Apprenticeship program covers roughly 1,000 career options, from electrician and pipefitter to dental assistant and law enforcement agent. Regions would be wise to support the development of more local apprenticeship programs that promote wide opportunities to earn, learn, and stay in local communities.

President Donald Trump, who spent 14 seasons hosting the reality game shows *The Apprentice* and *The Celebrity Apprentice*, has made apprenticeships a focus of his early tenure. Trump announced in June 2017 his plan to grow the number of apprenticeships nationwide to 5 million—a 10-fold increase—by 2022. However, with relatively little additional proposed funding and few details on implementation, the odds of that lofty goal becoming reality are low (Breuninger 2017).

5) Make sure that high school students receive career guidance, and support interactions between high school guidance counselors, teachers, and local employers, especially manufacturers.

Ask many manufacturers about the challenges they face in attracting new workers, and many of them will point to high school for the following reasons. 1) State assessments often directly or indirectly incentivize school districts to promote college attendance over direct entry into the workforce. 2) A decades-long decline in hands-on vocational training programs, especially in industrial arts, has disconnected high school learning from work readiness. 3) Guidance counselors and teachers have an outdated view of manufacturing that, combined with their own personal experience with college going, make them biased against promoting training and jobs in manufacturing.

Certainly, decades of job losses and declining relative wages (Ruckelshaus and Leberstein 2014) give teachers—and parents—good reason to be less than enthusiastic about promoting manufacturing career paths for young people. Yet, as the analysis presented here suggests, human capital needs are complicated and varied. Opportunities for good-paying, skilled work in manufacturing and the trades exist, and difficulties filling such positions occur.

Connecting young people to these opportunities may be a matter of educating the educators. For example, the Alliance for Working Together, a consortium of mostly small manufacturers in Northeast Ohio, has been proactively building relationships with area high schools and partnering in the development of a career academy promoting problem-based learning, flipped classrooms, and real-world experiences. Conexus Indiana, the initiative supporting advanced manufacturing and logistics, has recruited dozens of teachers and provided them with information about careers and skill needs so that they will serve as manufacturing "champions" within their Indiana high schools.

6) Recognize that skill development and ultimate career readiness begin early.

Much of the attention on the need for scientists, engineers, and other STEM workers is directed at the postsecondary level. However, creating more college graduates in STEM fields requires stepped-up STEM learning at the high school and primary levels. The five-year strategic plan for Federal Science, Technology, Engineering, and Mathematics (STEM) Education released in 2013 called for public-private partnerships aimed at developing 100,000 new STEM teachers to improve education from preschool to high school. Moreover, school districts throughout the country have added STEM programs, built STEM-specialty schools, and realigned curricula to emphasize STEM learning. However, as the analysis presented here makes clear, many of the high-skilled STEM jobs that are the focus of so much policy and media attention require more than knowledge of individual STEM domains. They also include high levels of creative thinking, complex problem solving, oral expression, reading comprehension, critical thinking, and information ordering. In other words, they also require a high level of "Soft" skills.

Most jobs—and many of the occupations associated with improved regional economic well-being—require far less than the level of STEM knowledge associated with higher education. In addition, workers at the highest end of the STEM skill spectrum tend to be relatively well-compensated for their abilities. These realities suggest that perhaps a more appropriate regional policy focus is on human capital development at the level below the bachelor's degree. All signs point to a need for experientially based learning modalities coupled with structured work experiences that can grow into apprentice-like opportunities.

Certainly, the increasing policy attention directed at the role of community colleges is warranted, but much of the challenge in developing a future workforce remains at the level of greatest local and regional responsibility—secondary and primary schools. Long-term regional economic development means ensuring that, by the end of high school, young people have the science knowledge, technical awareness, and numeracy skills—as well as the communication, thinking, and interpersonal skills—to succeed in a rapidly changing work environment. This type of high school success demands quality education in preschool and the primary grades, as well as exposure to career opportunities, as the critical building blocks to the skilled workforce of the future. This supports the importance of investing in early childhood education as an appropriate focus of long-term regional economic development initiatives based on human capital (Bartik 2011). It also argues for early work experiences that are connected to learning.

7) Pay attention to the bottom of the skills spectrum.

Much policy and media attention is focused at the top of the skills spectrum—the high-skilled "knowledge" workers seen as critical to the innovations and inventions that drive economic growth. As this research has demonstrated, however, regions vary in their demands for workers with such skills. They vary in their abilities to retain such talent and even in their abilities to hold onto the innovations developed within their bounds. (Remember, Facebook actually originated in Cambridge, Massachusetts, not Silicon Valley, and the first widely popular Web browser sprouted near the cornfields of Illinois.) Moreover, whatever boon in regional median wages or productivity that may come from regions' highly skilled workers is offset, or even eclipsed, by the drag on regional economic well-being associated with high concentrations of low-skill employment. Focusing simply on raising the educational attainment and skill levels of regional workers isn't going to change the reality that many regional jobs require very little skill.

8) Brace for change and disruption.

The rapid technological change that has wrought such sweeping changes to labor markets and economies will continue. Technological advancements looming in the not-too-distant future portend major disruptions in regional human capital demand, as well as human-capital-based interventions. For example, Google's self-driving car and Tesla's Autopilot software already available on models show that driverless vehicles are no longer the stuff of science fiction. A 2016 *Business Insider* report predicts the debut of fully autonomous vehicles in 2019 and 10 million self-driving cars on the road by 2020 (Greenough

2016). While promising improved safety on the roads and less stress in people's lives, the proliferation of autonomous vehicles will likely threaten a number of semi-skilled occupations—taxi drivers, couriers, truck drivers, and bus drivers, to name a few. This will have profound effects on employment, especially in regions that have high demand for such workers.

On the other end of the skills spectrum, technological advancements are likely to reshape even one of the key mechanisms by which human capital is developed—the educational system. Classrooms, from the primary grades to the master's level, are already being "flipped," where students watch video content at home and use class time to practice and interact. Distance learning and mobile online courses are enabling learners to access top educators—and increasingly receive a credential for it. "Open-source" and entrepreneurial educational systems are emerging to enable learners more control and to allow a new wave of content providers opportunities to share and profit from the knowledge and materials they create. These developments, especially in light of mounting costs related to more traditional educational settings, hold promise for better access to quality education but also augur potentially huge changes in educational employment.

CLOSING THOUGHTS

This is both an exciting and unsettling time for individuals and societies faced with making choices about which skills to invest in and what knowledge to pursue. The opportunities open to those workers and regions with the right mix of talent and luck are extraordinary. The speed with which technology is remaking work and demands for talent are equally breathtaking. This makes for a difficult milieu for strategic private and public investments in human capital development. Yet, technology that is disrupting the workplace can also facilitate better understanding of job demands and skill concentrations and enable cheaper, quicker, more accessible, and better targeted pathways to developing the human capital that rewards individuals and regions.

Appendix A
Occupational Wage and Employment
by High/Mid/Low STEM/Soft KSAs

(Total number of occupations = 764; 2014 U.S. Employment = 131,813,150)
*Median wage not reported for seven medical occupations; wages reflect level below median.

High STEM/High Soft			
No. of OCCs	Employment	No. of BA+ OCCs	Median Wage
95	9,759,870	79	$79,930
% of All OCCs	% of U.S. Emp	% OCCs BA+	Mean Wage
12.4%	7.4%	83.2%	$81,028
No. of OCCs Post HS	Post HS Emp	BA+ EMP	BA+ Med. Wage
15	3,842,510	5,718,750	$82,050
% of H/H OCCs Post HS	% Emp Post HS	% of H/H Emp BA+	Post HS Med. Wg
15.8%	39.4%	58.6%	$56,130

OCC Code	Occupation Description	Employment	Education	Median Wage
11-1021	General and Operations Managers	2,049,870	AD	$97,270
11-3021	Computer and Information Systems Managers	330,360	BA	$127,640
11-3051	Industrial Production Managers	167,200	AD	$92,470
11-3061	Purchasing Managers	70,840	BA	$106,090
11-9013	Farmers, Ranchers, and Other Agricultural Managers	4,300	BA	$68,050
11-9021	Construction Managers	227,710	BA	$85,630
11-9041	Architectural and Engineering Managers	179,320	BA	$130,620
11-9051	Food Service Managers	198,610	HS	$48,560
11-9081	Lodging Managers	31,740	Some College	$47,680
11-9111	Medical and Health Services Managers	310,320	BA	$92,810

153

11-9121	Natural Sciences Managers	53,290	MA	$120,050
11-9161	Emergency Management Directors	9,770	BA	$64,360
13-1041	Compliance Officers	246,970	BA	$64,950
13-1081	Logisticians	125,670	BA	$73,870
13-1199	Business Operations Specialists, All Other	934,370	BA	$67,280
15-1111	Computer and Information Research Scientists	24,210	MA	$108,360
15-1121	Computer Systems Analysts	528,320	AD	$82,710
15-1122	Information Security Analysts	80,180	BA	$88,890
17-1011	Architects, Except Landscape and Naval	88,900	BA	$74,520
17-1012	Landscape Architects	18,110	BA	$64,570
17-1022	Surveyors	41,970	BA	$57,050
17-2011	Aerospace Engineers	69,080	BA	$105,380
17-2021	Agricultural Engineers	2,450	BA	$71,730
17-2031	Biomedical Engineers	20,080	MA	$86,950
17-2041	Chemical Engineers	33,470	BA	$96,940
17-2051	Civil Engineers	263,460	BA Certificate	$82,050
17-2071	Electrical Engineers	174,550	BA	$91,410
17-2072	Electronics Engineers, Except Computer	133,990	BA	$95,790
17-2081	Environmental Engineers	53,240	MA	$83,360
17-2111	Health and Safety Engineers, Except Mining Safety Engineers . . .	24,530	BA	$81,830
17-2112	Industrial Engineers	236,990	BA	$81,490
17-2121	Marine Engineers and Naval Architects	7,570	BA	$92,930
17-2131	Materials Engineers	24,990	BA	$87,690
17-2141	Mechanical Engineers	270,700	BA	$83,060
17-2151	Mining and Geological Engineers, Including Mining Safety Engineers	8,200	BA	$90,160
17-2161	Nuclear Engineers	16,520	BA	$100,470
17-2171	Petroleum Engineers	33,740	BA	$130,050
17-2199	Engineers, All Other	124,570	BA	$94,240
19-1011	Animal Scientists	2,350	Doctoral	$61,110
19-1012	Food Scientists and Technologists	14,170	BA	$61,480
19-1013	Soil and Plant Scientists	15,150	Doctoral	$59,920
19-1021	Biochemists and Biophysicists	31,350	Doctoral	$84,940

19-1022	Microbiologists	20,670	BA Certificate	$67,790
19-1023	Zoologists and Wildlife Biologists	18,970	MA	$58,270
19-1029	Biological Scientists, All Other	32,230	Doctoral	$74,720
19-1031	Conservation Scientists	19,210	BA	$61,860
19-1032	Foresters	9,140	BA	$57,980
19-1042	Medical Scientists, Except Epidemiologists	100,740	Doctoral	$79,930
19-2012	Physicists	16,790	Doctoral	$109,600
19-2032	Materials Scientists	6,900	MA	$91,980
19-2041	Environmental Scientists and Specialists, Including Health	88,740	BA	$66,250
19-2042	Geoscientists, Except Hydrologists and Geographers	34,000	BA	$89,910
19-2043	Hydrologists	6,580	BA	$78,370
19-2099	Physical Scientists, All Other	23,030	MA	$94,030
19-3099	Social Scientists and Related Workers, All Other	32,010	BA	$75,630
19-4092	Forensic Science Technicians	13,570	AD	$55,360
25-1021	Computer Science Teachers, Postsecondary	35,410	MA	$72,010
25-1031	Architecture Teachers, Postsecondary	7,190	First Professional	$73,720
25-1032	Engineering Teachers, Postsecondary	36,650	Doctoral	$94,130
25-1043	Forestry and Conservation Science Teachers, Postsecondary	1,850	Doctoral	$84,090
25-1051	Atmospheric, Earth, Marine, and Space Sciences Teachers . . .	10,890	Doctoral	$81,780
25-1052	Chemistry Teachers, Postsecondary	21,470	Doctoral	$73,080
25-1054	Physics Teachers, Postsecondary	14,160	Doctoral	$80,720
25-2023	Career/Technical Education Teachers, Middle School	14,000	BA	$54,090
25-2032	Career/Technical Education Teachers, Secondary School	81,560	BA	$55,200
25-9021	Farm and Home Management Advisors	8,900	MA	$46,520
27-1025	Interior Designers	45,010	BA	$48,400
27-1027	Set and Exhibit Designers	10,460	BA Certificate	$49,810
29-1021	Dentists, General	97,990	Doctoral	$149,540

29-1022	Oral and Maxillofacial Surgeons	5,120	Post Doc	$155,740*
29-1023	Orthodontists	6,190	Post Doc	$118,290*
29-1031	Dietitians and Nutritionists	59,490	BA Certificate	$56,950
29-1041	Optometrists	33,340	Doctoral	$101,410
29-1067	Surgeons	41,070	Doctoral	$130,710*
29-1081	Podiatrists	8,910	Doctoral	$120,700
29-1128	Exercise Physiologists	6,660	MA	$46,270
29-1131	Veterinarians	62,470	Doctoral	$87,590
29-1151	Nurse Anesthetists	36,590	MA	$153,780
29-1171	Nurse Practitioners	122,050	MA	$95,350
29-1181	Audiologists	12,250	Doctoral	$73,060
29-2041	Emergency Medical Technicians and Paramedics	235,760	Post HS Certificate	$31,700
29-2091	Orthotists and Prosthetists	7,830	BA Certificate	$64,040
29-9011	Occupational Health and Safety Specialists	65,130	BA	$69,210
29-9012	Occupational Health and Safety Technicians	13,990	AD	$48,120
33-1021	First-Line Supervisors of Fire Fighting and Prevention Workers	59,870	Post HS Certificate	$70,670
33-2021	Fire Inspectors and Investigators	11,370	Some College	$56,130
33-3021	Detectives and Criminal Investigators	108,720	Post HS Certificate	$79,870
33-3031	Fish and Game Wardens	5,820	BA	$50,880
35-1011	Chefs and Head Cooks	118,130	AD	$41,610
39-4031	Morticians, Undertakers, and Funeral Directors	25,160	AD	$47,250
41-9031	Sales Engineers	68,080	BA	$96,340
43-9031	Desktop Publishers	13,310	AD	$38,200
49-1011	First-Line Supervisors of Mechanics, Installers, and Repairers	434,810	Post HS Certificate	$62,150
53-2011	Airline Pilots, Copilots, and Flight Engineers	75,760	BA	$118,140
53-5021	Captains, Mates, and Pilots of Water Vessels	30,690	Post HS Certificate	$72,340

High STEM/Mid Soft			
No. of OCCs	Employment	No. of BA+ OCCs	Median Wage
92	8,722,820	22	$50,785
% of All OCCs	% of U.S. Emp	% OCCs BA+	Mean Wage
12.0%	6.6%	23.9%	$53,752
No. of OCCs Post HS	Post HS Emp	BA+ Emp	BA+ Med. Wage
51	5,415,700	2,159,130	$52,525
% of H/M OCCs Post HS	% Emp Post HS	% of H/M Emp BA+	Post HS Med. Wg
55.4%	62.1%	24.8%	$52,200

OCC Code	Occupation	Employment	Education	Median Wage
13-1021	Buyers and Purchasing Agents, Farm Products	11,250	BA	$55,080
13-1051	Cost Estimators	209,130	BA	$60,050
15-1132	Software Developers, Applications	686,470	BA	$95,510
15-1141	Database Administrators	112,170	BA	$80,280
15-1142	Network and Computer Systems Administrators	365,430	BA	$75,790
15-1143	Computer Network Architects	140,080	BA	$98,430
15-1151	Computer User Support Specialists	563,540	AD	$47,610
17-2061	Computer Hardware Engineers	76,360	BA	$108,430
17-3011	Architectural and Civil Drafters	91,520	BA	$49,970
17-3013	Mechanical Drafters	64,070	AD	$52,200
17-3021	Aerospace Engineering and Operations Technicians	11,230	BA	$63,780
17-3023	Electrical and Electronics Engineering Technicians	137,040	AD	$59,820
17-3024	Electro-Mechanical Technicians	14,430	Post HS Certificate	$53,070
17-3025	Environmental Engineering Technicians	18,080	BA	$48,170
17-3026	Industrial Engineering Technicians	65,680	Some College	$53,370
17-3027	Mechanical Engineering Technicians	47,560	Some College	$53,530
17-3029	Engineering Technicians, Except Drafters, All Other	67,640	HS	$61,580

17-3031	Surveying and Mapping Technicians	50,750	Post HS Certificate	$40,770
19-2031	Chemists	85,970	BA	$73,480
19-4011	Agricultural and Food Science Technicians	20,640	BA Certificate	$35,140
19-4021	Biological Technicians	72,640	BA	$41,290
19-4031	Chemical Technicians	63,760	AD	$44,180
19-4041	Geological and Petroleum Technicians	16,020	Some College	$54,810
19-4051	Nuclear Technicians	6,380	Some College	$74,690
19-4091	Environmental Science and Protection Technicians, . . .	33,760	BA	$42,190
19-4093	Forest and Conservation Technicians	30,310	Some College	$35,260
19-4099	Life, Physical, and Social Science Technicians, All Other	67,140	AD	$44,650
25-4013	Museum Technicians and Conservators	9,950	BA Certificate	$39,940
25-9011	Audio-Visual and Multimedia Collections Specialists	8,960	BA	$44,070
27-1021	Commercial and Industrial Designers	29,410	BA	$64,620
27-4011	Audio and Video Equipment Technicians	60,200	Some College	$41,780
27-4012	Broadcast Technicians	26,600	Post HS Certificate	$36,560
27-4014	Sound Engineering Technicians	13,750	Post HS Certificate	$49,870
29-1024	Prosthodontists	630	Post Doc	$100,280
29-1124	Radiation Therapists	16,380	AD	$80,090
29-1126	Respiratory Therapists	119,410	AD	$56,730
29-2012	Medical and Clinical Laboratory Technicians	160,460	BA	$38,370
29-2032	Diagnostic Medical Sonographers	59,760	AD	$67,530
29-2033	Nuclear Medicine Technologists	20,320	AD	$72,100
29-2035	Magnetic Resonance Imaging Technologists	33,130	AD	$67,090
29-2054	Respiratory Therapy Technicians	10,610	AD	$47,810
29-2055	Surgical Technologists	98,450	Some College	$43,350

29-2099	Health Technologists and Technicians, All Other	96,170	AD	$41,420
33-2011	Firefighters	308,790	HS	$45,970
33-2022	Forest Fire Inspectors and Prevention Specialists	1,630	BA	$36,430
37-1012	First-Line Supervisors of Landscaping, Lawn Service . . .	101,190	HS	$43,160
37-2021	Pest Control Workers	67,640	HS	$30,660
45-1011	First-Line Supervisors of Farming, Fishing, and Forestry...	18,530	HS	$44,880
45-4011	Forest and Conservation Workers	6,870	BA	$27,160
47-1011	First-Line Supervisors of Construction Trades . . .	496,370	Post HS Certificate	$60,990
47-2111	Electricians	566,930	Post HS Certificate	$51,110
47-4011	Construction and Building Inspectors	88,410	Some College	$56,040
47-4021	Elevator Installers and Repairers	20,590	Post HS Certificate	$78,620
47-4071	Septic Tank Servicers and Sewer Pipe Cleaners	24,350	HS	$34,810
47-5012	Rotary Drill Operators, Oil and Gas	26,480	HS	$53,160
47-5013	Service Unit Operators, Oil, Gas, and Mining	62,080	HS	$44,970
47-5031	Explosives Workers, Ordnance Handling Experts . . .	7,970	HS	$52,140
49-2022	Telecommunications Equipment Installers . . .	213,620	Post HS Certificate	$55,190
49-2091	Avionics Technicians	17,150	AD	$56,910
49-2094	Electrical and Electronics Repairers, Commercial . . .	65,900	Some College	$54,640
49-2095	Electrical and Electronics Repairers, Powerhouse . . .	22,120	Post HS Certificate	$71,400
49-2096	Electronic Equipment Installers and Repairers . . .	11,460	Post HS Certificate	$31,020
49-2097	Electronic Home Entertainment Equipment Installers . . .	26,590	Post HS Certificate	$36,090
49-3011	Aircraft Mechanics and Service Technicians	116,830	Post HS Certificate	$56,990
49-3023	Automotive Service Technicians and Mechanics	633,390	Post HS Certificate	$37,120

49-3042	Mobile Heavy Equipment Mechanics, Except Engines	119,280	Post HS Certificate	$47,580
49-9021	Heating, Air Conditioning, and Refrigeration Mechanics . . .	261,390	Post HS Certificate	$44,630
49-9044	Millwrights	39,290	HS	$50,460
49-9051	Electrical Power-Line Installers and Repairers	114,540	Post HS Certificate	$65,930
49-9062	Medical Equipment Repairers	41,430	Post HS Certificate	$45,660
49-9081	Wind Turbine Service Technicians	3,710	Post HS Certificate	$48,800
49-9092	Commercial Divers	3,620	Post HS Certificate	$45,890
49-9095	Manufactured Building and Mobile Home Installers	3,280	Less than HS	$29,600
49-9099	Installation, Maintenance, and Repair Workers . . .	138,460	HS	$37,220
51-1011	First-Line Supervisors of Production and Operating . . .	592,830	Post HS Certificate	$55,520
51-4011	Computer-Controlled Machine Tool Operators, Metal . . .	148,040	Post HS Certificate	$36,440
51-4012	CNCN Machine Tool Programmers, Metal and Plastic	24,960	AD	$47,500
51-8011	Nuclear Power Reactor Operators	7,400	HS	$82,500
51-8012	Power Distributors and Dispatchers	11,180	HS	$78,240
51-8031	Water and Wastewater Treatment Plant and System . . .	111,640	HS	$44,100
51-8091	Chemical Plant and System Operators	37,490	HS	$55,900
51-8092	Gas Plant Operators	16,320	Post HS Certificate	$64,100
51-8093	Petroleum Pump System Operators, Refinery Operators . . .	41,700	HS	$62,830
51-9021	Crushing, Grinding, and Polishing Machine Setters . . .	29,980	HS	$33,070
51-9082	Medical Appliance Technicians	13,290	Some College	$35,580
53-2012	Commercial Pilots	38,170	Some College	$75,620
53-4031	Railroad Conductors and Yardmasters	42,900	HS	$54,770
53-5022	Motorboat Operators	4,060	Post HS Certificate	$37,120

53-5031	Ship Engineers	10,060	Post HS Certificate	$68,100
53-6041	Traffic Technicians	6,490	BA	$43,430
53-6051	Transportation Inspectors	24,350	Post HS Certificate	$69,170
53-7011	Conveyor Operators and Tenders	38,830	Some College	$31,220

High STEM/Low Soft

No. of OCCs	Employment	No. of BA+ OCCs	Median Wage
53	5,915,670	0	$39,100
% of All OCCs	% of U.S. Emp	% OCCs BA+	Mean Wage
6.9%	4.5%	–	$41,052
No. of OCCs Post HS	Post HS Emp	BA+ Emp	BA+ Med. Wage
11	931,290	0	–
% of H/L OCCs Post HS	% Emp Post HS	% of H/L Emp BA+	Post HS Med. Wg
20.8%	15.7%	–	$47,950

OCC Code	Occupation	Employment	Education	Median Wage
31-9093	Medical Equipment Preparers	50,550	Post HS Certificate	$32,260
45-3011	Fishers and Related Fishing Workers	400	HS	$35,250
47-2011	Boilermakers	17,210	HS	$59,860
47-2021	Brickmasons and Blockmasons	59,340	Less than HS	$47,650
47-2031	Carpenters	617,060	HS	$40,820
47-2061	Construction Laborers	852,870	HS	$31,090
47-2071	Paving, Surfacing, and Tamping Equipment Operators	54,940	HS	$38,660
47-2131	Insulation Workers, Floor, Ceiling, and Wall	24,180	Less than HS	$33,720
47-2152	Plumbers, Pipefitters, and Steamfitters	372,570	HS	$50,660
47-2211	Sheet Metal Workers	132,530	HS	$45,070
47-2221	Structural Iron and Steel Workers	60,010	HS	$48,200
47-2231	Solar Photovoltaic Installers	5,170	HS	$40,020
47-3012	Helpers–Carpenters	38,900	HS	$26,600
47-4051	Highway Maintenance Workers	140,650	HS	$36,580

47-4099	Construction and Related Workers, All Other	31,190	HS	$35,400
47-5011	Derrick Operators, Oil and Gas	20,760	Less than HS	$48,410
47-5041	Continuous Mining Machine Operators	11,540	Less than HS	$48,440
47-5071	Roustabouts, Oil and Gas	73,450	HS	$35,780
49-2021	Radio, Cellular, and Tower Equipment Installers and Repairers	13,310	Post HS Certificate	$47,950
49-2092	Electric Motor, Power Tool, and Related Repairers	17,380	HS	$39,220
49-2093	Electrical and Electronics Installers and Repairers, Transportation . . .	14,160	Post HS Certificate	$56,000
49-3031	Bus and Truck Mechanics and Diesel Engine Specialists	243,080	HS	$43,630
49-3041	Farm Equipment Mechanics and Service Technicians	35,320	HS	$36,150
49-3053	Outdoor Power Equipment and Other Small Engine Mechanics	29,220	HS	$32,120
49-3092	Recreational Vehicle Service Technicians	10,990	HS	$35,630
49-9012	Control and Valve Installers and Repairers, Except Mechanical Door	41,290	Post HS Certificate	$53,140
49-9041	Industrial Machinery Mechanics	313,880	Post HS Certificate	$48,630
49-9043	Maintenance Workers, Machinery	90,730	HS	$42,640
49-9052	Telecommunications Line Installers and Repairers	114,420	HS	$54,450
49-9061	Camera and Photographic Equipment Repairers	3,150	Post HS Certificate	$40,020
49-9071	Maintenance and Repair Workers, General	1,282,920	HS	$36,170
49-9094	Locksmiths and Safe Repairers	17,090	HS	$38,600
49-9097	Signal and Track Switch Repairers	7,880	Some College	$60,640
51-2023	Electromechanical Equipment Assemblers	46,990	HS	$32,760
51-4021	Extruding and Drawing Machine Setters, Operators, and Tenders . . .	72,520	HS	$32,610
51-4032	Drilling and Boring Machine Tool Setters, Operators, and Tenders . . .	17,470	Post HS Certificate	$34,500
51-4033	Grinding, Lapping, Polishing, and Buffing Machine Tool Setters . . .	70,130	HS	$32,660

51-4041	Machinists	392,700	Post HS Certificate	$39,980
51-4061	Model Makers, Metal and Plastic	6,140	HS	$46,180
51-4081	Multiple Machine Tool Setters, Operators, and Tenders . . .	98,160	HS	$34,140
51-4111	Tool and Die Makers	75,950	Post HS Certificate	$48,890
51-4191	Heat Treating Equipment Setters, Operators, and Tenders, Metal . . .	20,980	HS	$35,320
51-6061	Textile Bleaching and Dyeing Machine Operators and Tenders	11,510	HS	$24,930
51-7011	Cabinetmakers and Bench Carpenters	88,170	HS	$31,580
51-7032	Patternmakers, Wood	950	Post HS Certificate	$37,980
51-7042	Woodworking Machine Setters, Operators, and Tenders, Except...	70,810	HS	$27,450
51-8021	Stationary Engineers and Boiler Operators	37,550	HS	$56,330
51-8099	Plant and System Operators, All Other	11,610	HS	$55,230
51-9012	Separating, Filtering, Clarifying, Precipitating, and Still Machine...	43,310	HS	$38,590
51-9193	Cooling and Freezing Equipment Operators and Tenders	8,070	HS	$28,280
53-5011	Sailors and Marine Oilers	27,640	HS	$39,100
53-7071	Gas Compressor and Gas Pumping Station Operators	4,700	HS	$56,280
53-7072	Pump Operators, Except Wellhead Pumpers	12,170	HS	$43,500

Mid STEM/High Soft			
No. of OCCs	Employment	No. of BA+ OCCs	Median Wage
74	14,919,310	65	$72,845
% of All OCCs	% of U.S. Emp	% OCCs BA+	Mean Wage
9.7%	11.3%	87.8%	$81,581
No. of OCCs Post HS	Post HS Emp	BA+ Emp	BA+ Med. Wage
9	5,809,770	9,109,540	$74,750
% of M/H OCCs Post HS	% Emp Post HS	% of M/H Emp BA+	Post HS Med. Wg
12.2%	38.9%	61.1%	$66,640

OCC Code	Occupation	Employment	Education	Median Wage
11-1011	Chief Executives	246,240	MA	$173,320
11-2011	Advertising and Promotions Managers	29,340	BA	$96,720
11-2021	Marketing Managers	184,490	BA	$127,130
11-2022	Sales Managers	358,920	BA	$110,660
11-3071	Transportation, Storage, and Distribution Managers	106,000	BA	$85,400
11-3121	Human Resources Managers	116,610	BA	$102,780
11-3131	Training and Development Managers	29,870	BA	$101,930
11-9032	Education Administrators, Elementary and Secondary . . .	231,800	MA	$89,540
11-9039	Education Administrators, All Other	31,920	MA	$77,020
11-9071	Gaming Managers	3,870	Post HS Certificate	$67,310
11-9151	Social and Community Service Managers	116,670	BA	$62,740
11-9199	Managers, All Other	361,900	BA	$105,060
13-1023	Purchasing Agents, Except Wholesale, Retail, and Farm . . .	288,430	BA	$60,980
13-1111	Management Analysts	587,450	BA	$80,880
13-1121	Meeting, Convention, and Event Planners	77,940	BA	$46,490
13-1161	Market Research Analysts and Marketing Specialists	468,160	BA	$61,290
13-2099	Financial Specialists, All Other	137,460	BA	$65,440
19-1041	Epidemiologists	5,420	MA	$67,420
19-2011	Astronomers	1,660	Post Doc	$105,410
19-2021	Atmospheric and Space Scientists	10,850	BA	$87,980
19-3011	Economists	18,680	Doctoral Degree	$95,710
19-3022	Survey Researchers	15,410	MA	$49,760
19-3032	Industrial-Organizational Psychologists	1,110	Doctoral Degree	$76,950
19-3039	Psychologists, All Other	11,980	Post Doc	$92,110
19-3051	Urban and Regional Planners	35,820	MA	$66,940
19-3091	Anthropologists and Archeologists	7,040	MA	$59,280

19-3092	Geographers	1,260	MA	$76,420
21-2011	Clergy	46,510	BA	$43,950
25-1011	Business Teachers, Postsecondary	85,030	Doctoral Degree	$74,090
25-1041	Agricultural Sciences Teachers, Postsecondary	9,890	Doctoral Degree	$86,260
25-1042	Biological Science Teachers, Postsecondary	52,750	Doctoral Degree	$74,580
25-1053	Environmental Science Teachers, Postsecondary	5,300	Doctoral Degree	$77,470
25-1061	Anthropology and Archeology Teachers, Postsecondary	6,100	Doctoral Degree	$74,750
25-1064	Geography Teachers, Postsecondary	4,440	Doctoral Degree	$71,320
25-1066	Psychology Teachers, Postsecondary	37,930	Doctoral Degree	$68,690
25-1071	Health Specialties Teachers, Postsecondary	168,090	Doctoral Degree	$90,210
25-1072	Nursing Instructors and Teachers, Postsecondary	56,840	MA	$66,100
25-1111	Criminal Justice and Law Enforcement Teachers . . .	14,890	MA	$57,200
25-1192	Home Economics Teachers, Postsecondary	3,620	MA	$63,390
25-1193	Recreation and Fitness Studies Teachers, Postsecondary	18,650	MA	$58,280
25-2021	Elementary School Teachers, Except Special Education	1,353,020	BA	$54,120
25-2022	Middle School Teachers, Except Special & Career/Tech . . .	630,620	BA	$54,940
25-2031	Secondary School Teachers, Except Special & Career/Tech . . .	960,380	BA	$56,310
25-2054	Special Education Teachers, Secondary School	135,520	BA	$57,810
25-2059	Special Education Teachers, All Other	39,620	BA Certificate	$54,520
25-4012	Curators	11,200	MA	$51,280
25-9031	Instructional Coordinators	133,780	MA	$61,550
27-1011	Art Directors	33,140	BA	$85,610
27-2012	Producers and Directors	97,300	BA	$69,100
27-2022	Coaches and Scouts	211,760	BA	$30,640

29-1011	Chiropractors	29,830	Doctoral Degree	$66,720
29-1051	Pharmacists	290,780	First Professional	$120,950
29-1061	Anesthesiologists	30,060	Post Doc	$151,450*
29-1062	Family and General Practitioners	124,810	Doctoral Degree	$180,180
29-1063	Internists, General	48,390	Post Doc	$125,230*
29-1064	Obstetricians and Gynecologists	21,740	Doctoral Degree	$156,730*
29-1065	Pediatricians, General	31,010	Doctoral Degree	$163,350
29-1066	Psychiatrists	25,080	Post Doc	$181,880
29-1069	Physicians and Surgeons, All Other	311,320	Post Doc	$111,630*
29-1071	Physician Assistants	91,670	MA	$95,820
29-1122	Occupational Therapists	110,520	MA	$78,810
29-1123	Physical Therapists	200,670	MA	$82,390
29-1127	Speech-Language Pathologists	126,500	MA	$71,550
29-1141	Registered Nurses	2,687,310	AD	$66,640
29-1161	Nurse Midwives	5,110	MA	$96,970
29-2061	Licensed Practical and Licensed Vocational Nurses	695,610	Some College	$42,490
29-9091	Athletic Trainers	22,400	MA	$43,370
29-9099	Healthcare Practitioners and Technical Workers...	40,840	BA	$49,430
33-1012	First-Line Supervisors of Police and Detectives	101,420	AD	$80,930
33-3051	Police and Sheriff's Patrol Officers	638,810	Some College	$56,810
41-1012	First-Line Supervisors of Non-Retail Sales Workers	248,770	AD	$71,600
43-1011	First-Line Supervisors of Office and Administrative . . .	1,404,070	AD	$50,780
53-2021	Air Traffic Controllers	22,860	Post HS Certificate	$122,340
53-2022	Airfield Operations Specialists	7,050	AD	$49,180

	Mid STEM/Mid Soft			
	No. of OCCs	Employment	No. of BA+ OCCs	Median Wage
	53	7,687,390	12	$46,690
	% of All OCCs	% of U.S. Emp	% M/M OCCs BA+	Mean Wage
	6.9%	5.8%	22.6%	$50,331
	No. of M/M OCCs Post HS	Post HS Emp	BA+ Emp	BA+ Med. Wage
	25	2,565,010	1,534,060	$68,565
	% of M/M OCCs Post HS	% Emp Post HS	% of M/M Emp BA+	Post HS Med. Wg
	47.2%	33.4%	20.0%	$49,080

OCC Code	Occupation	Employment	Education	Median Wage
11-9061	Funeral Service Managers	8,330	AD	$68,870
11-9131	Postmasters and Mail Superintendents	17,930	HS	$65,800
13-1022	Wholesale and Retail Buyers, Except Farm Products	110,560	Some College	$52,270
13-1032	Insurance Appraisers, Auto Damage	13,690	AD	$63,420
13-2021	Appraisers and Assessors of Real Estate	63,220	AD	$52,570
15-1131	Computer Programmers	302,150	BA	$77,550
15-1133	Software Developers, Systems Software	382,400	BA	$102,880
15-1134	Web Developers	121,020	AD	$63,490
15-1152	Computer Network Support Specialists	174,490	BA	$61,830
15-1199	Computer Occupations, All Other	212,510	BA	$83,410
15-2021	Mathematicians	3,130	MA	$103,720
15-2031	Operations Research Analysts	86,950	MA	$76,660
17-1021	Cartographers and Photogrammetrists	11,610	BA	$60,930
17-3022	Civil Engineering Technicians	71,300	Post HS Certificate	$48,340
19-1020	Biological Scientists	103,210	MA	$71,940
19-4061	Social Science Research Assistants	27,780	BA	$39,460
25-1022	Mathematical Science Teachers, Postsecondary	54,010	Doctoral Degree	$65,190

27-4021	Photographers	52,250	Some College	$30,490
27-4031	Camera Operators, Television, Video, and Motion Picture	18,310	AD	$48,080
29-2011	Medical and Clinical Laboratory Technologists	161,710	BA	$59,430
29-2031	Cardiovascular Technologists and Technicians	51,080	AD	$54,330
29-2034	Radiologic Technologists	193,400	AD	$55,870
29-2056	Veterinary Technologists and Technicians	93,300	AD	$31,070
29-2057	Ophthalmic Medical Technicians	36,470	Post HS Certificate	$35,230
29-2081	Opticians, Dispensing	73,110	Post HS Certificate	$34,280
29-2092	Hearing Aid Specialists	5,570	AD	$43,010
31-2011	Occupational Therapy Assistants	32,230	AD	$56,950
31-2021	Physical Therapist Assistants	76,910	AD	$54,410
31-9092	Medical Assistants	584,970	Post HS Certificate	$29,960
31-9097	Phlebotomists	111,950	Post HS Certificate	$30,670
33-3012	Correctional Officers and Jailers	434,420	HS	$39,780
33-9099	Protective Service Workers, All Other	113,020	HS	$28,440
35-1012	First-Line Supervisors of Food Preparation and Serving Workers	867,340	HS	$29,560
37-1011	First-Line Supervisors of House-keeping and Janitorial Workers	168,960	HS	$36,270
39-1021	First-Line Supervisors of Personal Service Workers	161,990	Some College	$35,250
39-4011	Embalmers	3,650	AD	$41,720
39-5091	Makeup Artists, Theatrical and Performance	2,610	HS	$44,310
41-1011	First-Line Supervisors of Retail Sales Workers	1,199,770	HS	$37,860
41-4011	Sales Reps, Wholesale & Manufac-turing, Technical & Scientific	335,540	Some College	$75,140
41-9022	Real Estate Sales Agents	157,660	HS	$40,990
43-5031	Police, Fire, and Ambulance Dispatchers	96,390	HS	$37,410

OCC Code	Occupation	Employment	Education	Median Wage
43-5032	Dispatchers, Except Police, Fire, and Ambulance	190,330	HS	$36,690
43-9011	Computer Operators	58,060	Some College	$39,590
43-9111	Statistical Assistants	14,110	BA	$42,070
45-2011	Agricultural Inspectors	13,800	HS	$43,090
47-4041	Hazardous Materials Removal Workers	42,250	HS	$38,520
49-2011	Computer, Automated Teller, and Office Machine Repairers	110,940	Some College	$36,560
51-6092	Fabric and Apparel Patternmakers	5,440	Post HS Certificate	$41,310
51-9011	Chemical Equipment Operators and Tenders	64,710	HS	$48,090
53-1011	Aircraft Cargo Handling Supervisors	5,750	HS	$47,760
53-1021	First-Line Supervisors of Helpers, Laborers & Material Movers, Hand	171,720	Post HS Certificate	$46,690
53-1031	First-Line Supervisors, Transportation & Material-Moving Machine . . .	197,000	HS	$54,930
53-6061	Transportation Attendants, Except Flight Attendants	16,380	HS	$23,380

Mid STEM/Low Soft

No. of OCCs	Employment	No. of BA+ OCCs	Median Wage
105	6,630,800	1	$35,420
% of All OCCs	% of U.S. Emp	% OCCs BA+	Mean Wage
13.7%	5.0%	1.0%	$36,759
No. of M/L OCCs Post HS	Post HS Emp	BA+ Emp	BA+ Med. Wage
11	538,440	1,060	$54,140
% of M/L OCCs Post HS	% Emp Post HS	% of M/L Emp BA+	Post HS Med Wg
10.5%	8.1%	0.0%	$37,340

OCC Code	Occupation	Employment	Education	Median Wage
15-2091	Mathematical Technicians	1,060	BA	$54,140
17-3012	Electrical and Electronics Drafters	29,390	AD	$58,790
27-1012	Craft Artists	4,760	HS	$31,080
27-1013	Fine Artists, Including Painters, Sculptors, and Illustrators	12,100	Some College	$43,890

37-3012	Pesticide Handlers, Sprayers, and Applicators, Vegetation	23,790	Post HS Certificate	$31,240
37-3013	Tree Trimmers and Pruners	39,640	HS	$32,960
39-3021	Motion Picture Projectionists	6,290	HS	$20,830
43-9071	Office Machine Operators, Except Computer	66,530	HS	$28,510
45-2021	Animal Breeders	1,110	Post HS Certificate	$40,000
45-2091	Agricultural Equipment Operators	26,100	HS	$26,910
45-2092	Farmworkers & Laborers, Crop, Nursery, & Greenhouse	269,650	HS	$19,060
45-2093	Farmworkers, Farm, Ranch, and Aquacultural Animals	31,540	Less than HS	$22,930
45-4021	Fallers	6,090	Less than HS	$34,490
45-4022	Logging Equipment Operators	26,010	Less than HS	$35,190
47-2022	Stonemasons	11,250	HS	$37,880
47-2041	Carpet Installers	26,050	HS	$35,880
47-2044	Tile and Marble Setters	31,590	Less than HS	$38,980
47-2051	Cement Masons and Concrete Finishers	152,570	Less than HS	$36,760
47-2053	Terrazzo Workers and Finishers	3,250	HS	$39,090
47-2072	Pile-Driver Operators	3,470	HS	$51,510
47-2073	Operating Engineers & Other Construction Equipment . . .	344,510	HS	$43,510
47-2081	Drywall and Ceiling Tile Installers	85,020	HS	$38,100
47-2132	Insulation Workers, Mechanical	28,660	HS	$42,990
47-2141	Painters, Construction and Maintenance	204,600	HS	$35,950
47-2142	Paperhangers	3,570	HS	$32,930
47-2151	Pipelayers	41,080	Less than HS	$37,000
47-2171	Reinforcing Iron and Rebar Workers	18,530	HS	$50,020
47-2181	Roofers	103,650	HS	$35,760
47-3011	Helpers–Brickmasons, Blockmasons, Stonemasons . . .	23,570	HS	$28,830
47-3013	Helpers–Electricians	68,280	HS	$27,940
47-3016	Helpers–Roofers	11,640	HS	$26,060
47-4031	Fence Erectors	20,990	HS	$31,510
47-4061	Rail-Track Laying and Maintenance Equipment Operators	14,820	HS	$51,840

47-4091	Segmental Pavers	1,130	HS	$32,180
47-5021	Earth Drillers, Except Oil and Gas	19,160	HS	$43,540
47-5042	Mine Cutting and Channeling Machine Operators	6,960	HS	$50,260
47-5051	Rock Splitters, Quarry	3,630	HS	$33,240
47-5061	Roof Bolters, Mining	5,710	HS	$54,860
47-5081	Helpers–Extraction Workers	24,130	HS	$34,480
49-2098	Security and Fire Alarm Systems Installers	60,160	HS	$42,560
49-3021	Automotive Body and Related Repairers	137,140	HS	$40,320
49-3043	Rail Car Repairers	20,080	HS	$54,020
49-3051	Motorboat Mechanics and Service Technicians	20,210	Post HS Certificate	$37,340
49-3052	Motorcycle Mechanics	15,420	Post HS Certificate	$34,010
49-3091	Bicycle Repairers	10,520	HS	$26,370
49-9011	Mechanical Door Repairers	17,220	HS	$37,080
49-9031	Home Appliance Repairers	33,270	Post HS Certificate	$35,410
49-9045	Refractory Materials Repairers, Except Brickmasons	1,730	HS	$44,910
49-9063	Musical Instrument Repairers and Tuners	7,660	Post HS Certificate	$33,150
49-9091	Coin, Vending, and Amusement Machine Servicers . . .	30,840	HS	$31,860
49-9096	Riggers	20,350	HS	$41,570
49-9098	Helpers–Installation, Maintenance, and Repair Workers	126,980	HS	$25,390
51-2011	Aircraft Structure, Surfaces, Rigging & Systems Assemblers	40,630	HS	$48,340
51-2031	Engine and Other Machine Assemblers	38,330	HS	$38,310
51-2041	Structural Metal Fabricators and Fitters	78,050	HS	$36,570
51-2091	Fiberglass Laminators and Fabricators	18,770	HS	$28,950
51-2093	Timing Device Assemblers and Adjusters	1,650	HS	$30,060
51-4022	Forging Machine Setters, Operators, and Tenders, Metal...	21,340	HS	$33,710

51-4023	Rolling Machine Setters, Operators, and Tenders, Metal . . .	33,370	HS	$39,900
51-4031	Cutting, Punching, & Press Machine Setters, Operators . . .	190,250	HS	$30,680
51-4034	Lathe and Turning Machine Tool Setters, Operators . . .	42,570	HS	$36,260
51-4035	Milling and Planing Machine Setters, Operators & Tenders . . .	22,110	Post HS Certificate	$37,100
51-4051	Metal-Refining Furnace Operators and Tenders	20,850	HS	$41,140
51-4062	Patternmakers, Metal and Plastic	3,770	Post HS Certificate	$41,390
51-4121	Welders, Cutters, Solderers, and Brazers	369,610	Post HS Certificate	$37,420
51-4122	Welding, Soldering & Brazing Machine Setters, Operators...	55,360	HS	$35,180
51-4192	Layout Workers, Metal and Plastic	13,070	HS	$45,020
51-4194	Tool Grinders, Filers, and Sharpeners	10,860	HS	$35,420
51-5112	Printing Press Operators	166,750	HS	$35,100
51-5113	Print Binding and Finishing Workers	51,430	HS	$29,500
51-6042	Shoe Machine Operators and Tenders	3,550	HS	$24,750
51-6052	Tailors, Dressmakers, and Custom Sewers	20,200	Less than HS	$26,460
51-6062	Textile Cutting Machine Setters, Operators, and Tenders	14,370	HS	$25,590
51-6091	Extruding and Forming Machine Setters, Operators . . .	19,770	HS	$32,970
51-6093	Upholsterers	29,770	HS	$31,890
51-7031	Model Makers, Wood	1,360	HS	$30,940
51-8013	Power Plant Operators	40,300	HS	$70,070
51-9022	Grinding and Polishing Workers, Hand	29,320	HS	$28,340
51-9023	Mixing and Blending Machine Setters, Operators . . .	122,670	HS	$34,340
51-9032	Cutting and Slicing Machine Setters, Operators . . .	62,570	HS	$32,040
51-9041	Extruding, Forming, Pressing, and Compacting Machine . . .	67,490	HS	$32,100

51-9071	Jewelers and Precious Stone and Metal Workers	23,200	HS	$36,870
51-9081	Dental Laboratory Technicians	35,320	HS	$36,830
51-9083	Ophthalmic Laboratory Technicians	27,610	HS	$28,890
51-9121	Coating, Painting, and Spraying Machine Setters . . .	90,590	HS	$31,460
51-9141	Semiconductor Processors	23,580	HS	$34,680
51-9151	Photographic Process Workers and Processing Machine . . .	28,800	HS	$24,600
51-9191	Adhesive Bonding Machine Operators and Tenders	18,210	HS	$31,340
51-9194	Etchers and Engravers	8,630	HS	$29,250
51-9195	Molders, Shapers, and Casters, Except Metal and Plastic	34,610	HS	$29,820
51-9196	Paper Goods Machine Setters, Operators, and Tenders	92,170	HS	$35,260
51-9199	Production Workers, All Other	217,500	HS	$28,260
53-3011	Ambulance Drivers and Attendants, Except EMTs	19,350	HS	$24,080
53-3032	Heavy and Tractor-Trailer Truck Drivers	1,625,290	HS	$39,520
53-4011	Locomotive Engineers	38,470	HS	$54,500
53-4012	Locomotive Firers	1,610	HS	$46,740
53-4013	Rail Yard Engineers, Dinkey Operators, and Hostlers	3,900	HS	$43,880
53-6011	Bridge and Lock Tenders	3,280	Less than HS	$48,120
53-7021	Crane and Tower Operators	44,540	HS	$50,720
53-7031	Dredge Operators	1,900	HS	$40,950
53-7032	Excavating and Loading Machine and Dragline Operators	47,470	HS	$39,830
53-7033	Loading Machine Operators, Underground Mining	4,220	HS	$50,290
53-7061	Cleaners of Vehicles and Equipment	321,740	HS	$20,670
53-7073	Wellhead Pumpers	12,720	HS	$47,340
53-7121	Tank Car, Truck, and Ship Loaders	12,490	HS	$41,180

	Low STEM/High Soft		
No. of OCCs	Employment	No. of BA+ OCCs	Median Wage
53	5,278,770	51	$61,450
% of All OCCs	% of U.S. Emp	% OCCs BA+	Mean Wage
6.9%	4.0%	96.2%	$65,305
No. of OCCs Post HS	Post HS Emp	BA+ Emp	BA+ Med. Wage
1	6,030	5,227,590	$62,220
% of L/H OCCs Post HS	% Emp Post HS	% of L/H Emp BA+	Post HS Med Wg
1.9%	0.1%	99.0%	$44,250

OCC Code	Occupation	Employment	Education	Median Wage
11-2031	Public Relations and Fundraising Managers	56,920	BA	$101,510
11-3031	Financial Managers	518,030	BA Certificate	$115,320
11-9031	Education Administrators, Preschool and Childcare Center/Program	47,150	BA	$45,260
11-9033	Education Administrators, Postsecondary	131,070	MA	$88,390
13-1031	Claims Adjusters, Examiners, and Investigators	266,280	BA	$62,220
13-1131	Fundraisers	55,230	BA	$52,430
13-1151	Training and Development Specialists	239,500	BA	$57,340
13-2051	Financial Analysts	262,610	BA	$78,620
13-2052	Personal Financial Advisors	196,490	BA	$81,060
13-2061	Financial Examiners	36,830	BA	$76,310
15-2011	Actuaries	21,490	BA	$96,700
19-3031	Clinical, Counseling, and School Psychologists	104,730	Doctoral Degree	$68,900
19-3041	Sociologists	2,240	Doctoral Degree	$72,810
19-3094	Political Scientists	5,640	Doctoral Degree	$104,920
21-1011	Substance Abuse and Behavioral Disorder Counselors	85,180	MA	$39,270
21-1012	Educational, Guidance, School, and Vocational Counselors	246,280	MA	$53,370

21-1013	Marriage and Family Therapists	30,150	MA	$48,040
21-1014	Mental Health Counselors	120,010	MA	$40,850
21-1015	Rehabilitation Counselors	103,890	MA	$34,380
21-1021	Child, Family, and School Social Workers	286,520	BA	$42,120
21-1022	Healthcare Social Workers	145,920	MA	$51,930
21-1023	Mental Health and Substance Abuse Social Workers	109,460	MA	$41,380
21-1091	Health Educators	57,020	BA	$50,430
21-1094	Community Health Workers	47,880	BA	$34,870
21-2021	Directors, Religious Activities and Education	18,850	BA	$38,480
23-1011	Lawyers	603,310	Doctoral Degree	$114,970
23-1021	Administrative Law Judges, Adjudicators, and Hearing Officers	14,140	First Professional	$87,980
23-1023	Judges, Magistrate Judges, and Magistrates	28,090	Doctoral Degree	$115,140
25-1062	Area, Ethnic, and Cultural Studies Teachers, Postsecondary	9,150	Doctoral Degree	$68,950
25-1063	Economics Teachers, Postsecondary	13,710	Doctoral Degree	$90,870
25-1065	Political Science Teachers, Postsecondary	17,050	Doctoral Degree	$73,790
25-1067	Sociology Teachers, Postsecondary	16,900	Doctoral Degree	$67,880
25-1081	Education Teachers, Postsecondary	59,980	Doctoral Degree	$59,720
25-1082	Library Science Teachers, Postsecondary	4,540	Doctoral Degree	$66,580
25-1112	Law Teachers, Postsecondary	15,990	Doctoral Degree	$109,980
25-1113	Social Work Teachers, Postsecondary	10,970	Doctoral Degree	$62,440
25-1121	Art, Drama, and Music Teachers, Postsecondary	97,500	MA	$64,300
25-1122	Communications Teachers, Postsecondary	29,470	MA	$62,550
25-1123	English Language and Literature Teachers, Postsecondary	76,320	First Professional	$60,160

OCC Code	Occupation	Employment	Education	Median Wage
25-1125	History Teachers, Postsecondary	23,640	Doctoral Degree	$66,840
25-2053	Special Education Teachers, Middle School	94,820	BA Certificate	$56,760
25-3011	Adult Basic and Secondary Education and Literacy Teachers…	65,990	MA	$49,590
25-4021	Librarians	133,150	MA	$56,170
27-2032	Choreographers	6,030	Some College	$44,250
27-2041	Music Directors and Composers	21,880	BA	$48,180
27-3021	Broadcast News Analysts	4,310	BA	$61,450
27-3031	Public Relations Specialists	208,030	BA	$55,680
27-3041	Editors	97,350	BA	$54,890
27-3091	Interpreters and Translators	49,460	BA	$43,590
29-1125	Recreational Therapists	17,950	BA	$44,000
29-9092	Genetic Counselors	2,180	MA	$67,500
33-1011	First-Line Supervisors of Correctional Officers	45,150	HS	$57,970
41-3031	Securities, Commodities, and Financial Services Sales Agents	316,340	BA	$72,070

Low STEM/Mid Soft

No. of OCCs	Employment	No. of BA+ OCCs	Median Wage
98	27,867,710	31	$41,745
% of All OCCs	% of U.S. Emp	% OCCs BA+	Mean Wage
12.8%	21.1%	31.6%	$43,589
No. of OCCs Post HS	Post HS Emp	BA+ Emp	BA+ Med. Wage
38	10,242,510	5,871,440	$55,020
% of L/M OCCs Post HS	% Emp Post HS	% of L/M Emp BA+	Post HS Med. Wg
38.8%	36.8%	21.1%	$42,200

OCC Code	Occupation	Employment	Education	Median Wage
11-3011	Administrative Services Managers	268,730	AD	$83,790
11-3111	Compensation and Benefits Managers	16,380	BA	$108,070
11-9141	Property, Real Estate, and Community Association Managers	171,140	BA	$54,270

13-1011	Agents and Business Managers of Artists, Performers, and Athletes	11,860	BA	$64,200
13-1071	Human Resources Specialists	456,170	AD	$57,420
13-1141	Compensation, Benefits, and Job Analysis Specialists	80,970	BA	$60,600
13-2011	Accountants and Auditors	1,187,310	BA	$65,940
13-2031	Budget Analysts	57,120	BA	$71,220
13-2041	Credit Analysts	69,390	BA	$67,020
13-2053	Insurance Underwriters	91,720	BA	$64,220
13-2071	Credit Counselors	29,600	AD	$42,110
13-2072	Loan Officers	300,580	Some College	$62,620
13-2081	Tax Examiners and Collectors, and Revenue Agents	63,640	AD	$51,120
13-2082	Tax Preparers	68,590	AD	$35,990
15-2041	Statisticians	26,970	MA	$79,990
19-3093	Historians	3,220	MA Certificate	$55,870
21-1092	Probation Officers and Correctional Treatment Specialists	86,810	BA	$49,060
21-1093	Social and Human Service Assistants	354,800	BA	$29,790
23-1012	Judicial Law Clerks	11,660	Doctoral Degree	$48,640
23-1022	Arbitrators, Mediators, and Conciliators	6,710	BA Certificate	$57,180
23-2011	Paralegals and Legal Assistants	272,580	AD	$48,350
23-2093	Title Examiners, Abstractors, and Searchers	52,960	HS	$43,080
25-1124	Foreign Language and Literature Teachers, Postsecondary	30,880	MA	$59,490
25-1126	Philosophy and Religion Teachers, Postsecondary	23,210	Doctoral Degree	$63,630
25-1191	Graduate Teaching Assistants	126,030	MA	$31,570
25-1194	Vocational Education Teachers, Postsecondary	121,200	Post HS Certificate	$48,360
25-2011	Preschool Teachers, Except Special Education	352,420	Some College	$28,120
25-2012	Kindergarten Teachers, Except Special Education	158,240	BA	$50,600
25-3021	Self-Enrichment Education Teachers	202,360	BA	$36,020
25-4011	Archivists	5,360	MA	$49,120

25-9041	Teacher Assistants	1,192,590	Some College	$24,430
27-1014	Multimedia Artists and Animators	29,000	BA	$63,630
27-1022	Fashion Designers	17,840	AD	$64,030
27-1023	Floral Designers	45,050	HS	$24,750
27-1024	Graphic Designers	197,540	BA	$45,900
27-2021	Athletes and Sports Competitors	11,520	Less than HS	$43,350
27-2023	Umpires, Referees, and Other Sports Officials	17,510	HS	$24,090
27-3011	Radio and Television Announcers	30,220	Some College	$29,790
27-3012	Public Address System and Other Announcers	7,450	HS	$25,730
27-3022	Reporters and Correspondents	42,280	BA	$36,000
27-3042	Technical Writers	48,210	AD	$69,030
27-3043	Writers and Authors	43,500	BA	$58,850
27-4013	Radio Operators	1,100	HS	$46,380
27-4032	Film and Video Editors	24,460	AD	$57,210
29-1199	Health Diagnosing and Treating Practitioners, All Other	35,310	MA	$73,400
29-2021	Dental Hygienists	196,520	AD	$71,520
29-2051	Dietetic Technicians	28,690	HS	$25,780
29-2052	Pharmacy Technicians	368,760	HS	$29,810
29-2053	Psychiatric Technicians	64,540	BA	$31,130
31-1011	Home Health Aides	799,080	HS	$21,380
31-1013	Psychiatric Aides	72,860	HS	$26,220
31-2012	Occupational Therapy Aides	8,570	Some College	$26,550
31-9091	Dental Assistants	314,330	Post HS Certificate	$35,390
31-9099	Healthcare Support Workers, All Other	98,980	BA	$34,620
33-3052	Transit and Railroad Police	3,380	AD	$51,690
33-9011	Animal Control Workers	13,450	Post HS Certificate	$32,560
33-9021	Private Detectives and Investigators	26,880	Some College	$44,570
33-9031	Gaming Surveillance Officers and Gaming Investigators	10,030	Post HS Certificate	$29,840
33-9093	Transportation Security Screeners	43,220	HS	$38,090
35-3011	Bartenders	579,700	HS	$19,050
39-1011	Gaming Supervisors	24,100	HS	$49,420

39-2011	Animal Trainers	11,170	HS	$25,770
39-6012	Concierges	31,050	Some College	$28,170
39-7012	Travel Guides	3,090	Some College	$35,100
39-9011	Childcare Workers	582,970	HS	$19,730
39-9031	Fitness Trainers and Aerobics Instructors	241,000	Some College	$34,980
39-9032	Recreation Workers	321,110	BA	$22,620
39-9041	Residential Advisors	95,750	BA	$24,340
41-2022	Parts Salespersons	231,240	HS	$29,440
41-2031	Retail Salespersons	4,562,160	HS	$21,390
41-3011	Advertising Sales Agents	154,220	AD	$47,890
41-3021	Insurance Sales Agents	374,700	AD	$47,860
41-3041	Travel Agents	64,750	Post HS Certificate	$34,800
41-3099	Sales Representatives, Services, All Other	826,650	BA	$51,670
41-4012	Sales Reps, Wholesale & Manufacturing, Except Technical & Scientific Products	1,394,640	BA	$55,020
41-9021	Real Estate Brokers	38,720	Some College	$57,360
43-3011	Bill and Account Collectors	346,960	HS	$33,700
43-3031	Bookkeeping, Accounting, and Auditing Clerks	1,575,060	Some College	$36,430
43-3061	Procurement Clerks	70,190	Post HS Certificate	$39,930
43-3071	Tellers	514,520	HS	$25,760
43-4011	Brokerage Clerks	57,240	AD	$47,520
43-4021	Correspondence Clerks	7,580	HS	$35,460
43-4051	Customer Service Representatives	2,511,130	HS	$31,200
43-4061	Eligibility Interviewers, Government Programs	122,400	HS	$42,200
43-4071	File Clerks	148,280	Post HS Certificate	$27,580
43-4081	Hotel, Motel, and Resort Desk Clerks	241,140	HS	$20,610
43-4121	Library Assistants, Clerical	100,800	HS	$23,910
43-4131	Loan Interviewers and Clerks	212,440	Some College	$36,880
43-4141	New Accounts Clerks	52,260	HS	$34,000
43-4151	Order Clerks	190,390	HS	$31,180

43-4161	Human Resources Assistants, Except Payroll and Timekeeping	135,270	Some College	$38,040
43-4181	Reservation and Transportation Ticket Agents and Travel Clerks	138,260	HS	$33,510
43-5011	Cargo and Freight Agents	77,480	HS	$41,380
43-5061	Production, Planning, and Expediting Clerks	297,050	Post HS Certificate	$45,670
43-6011	Executive Secretaries and Executive Administrative Assistants	713,730	Some College	$51,270
43-6014	Secretaries & Administrative Assistants, Except Legal, Medical, Executive	2,207,220	Some College	$33,240
53-2031	Flight Attendants	98,510	Some College	$42,290
53-4041	Subway and Streetcar Operators	11,300	HS	$62,130

Low STEM/Low Soft			
No. of OCCs	Employment	No. of BA+ OCCs	Median Wage
141	45,030,810	2	$26,640
% of All OCCs	% of U.S. Emp	% OCCs BA+	Mean Wage
18.4%	34.2%	1.4%	$28,648
No. of OCCs Post HS	Post HS Emp	BA+ Emp	BA+ Med Wage
16	1,811,020	104,760	$33,330
% of L/L OCCs Post HS	% Emp Post HS	% of L/L Emp	Post HS Med Wg
11.3%	4.0%	0.2%	$33,495

OCC Code	Occupation	Employment	Education	Median Wage
13-1074	Farm Labor Contractors	950	HS	$41,110
23-2091	Court Reporters	18,330	Some College	$49,860
25-4031	Library Technicians	94,260	BA	$31,680
27-1026	Merchandise Displayers and Window Trimmers	93,000	HS	$26,590
29-2071	Medical Records and Health Information Technicians	184,740	Post HS Certificate	$35,900
31-1014	Nursing Assistants	1,427,740	HS	$25,100
31-2022	Physical Therapist Aides	48,730	HS	$24,650
31-9011	Massage Therapists	87,670	Post HS Certificate	$37,180

31-9094	Medical Transcriptionists	61,210	Post HS Certificate	$34,750
31-9095	Pharmacy Aides	41,240	HS	$23,200
31-9096	Veterinary Assistants and Laboratory Animal Caretakers	71,060	Post HS Certificate	$23,790
33-3011	Bailiffs	16,310	HS	$38,150
33-3041	Parking Enforcement Workers	8,680	Post HS Certificate	$36,570
33-9032	Security Guards	1,077,520	HS	$24,410
33-9091	Crossing Guards	66,310	HS	$24,750
33-9092	Lifeguards, Ski Patrol, & Other Recreational Protective . . .	135,070	HS	$19,090
35-2011	Cooks, Fast Food	519,910	Less than HS	$18,540
35-2012	Cooks, Institution and Cafeteria	402,800	HS	$23,440
35-2013	Cooks, Private Household	560	Post HS Certificate	$22,940
35-2014	Cooks, Restaurant	1,104,790	HS	$22,490
35-2015	Cooks, Short Order	180,800	HS	$20,190
35-2021	Food Preparation Workers	850,220	Less than HS	$19,560
35-3021	Combined Food Preparation and Serving Workers . . .	3,131,390	Less than HS	$18,410
35-3022	Counter Attendants, Cafeteria, Food Concession . . .	476,470	Less than HS	$18,740
35-3031	Waiters and Waitresses	2,445,230	HS	$18,730
35-3041	Food Servers, Nonrestaurant	250,840	Less than HS	$19,900
35-9011	Dining Room & Cafeteria Attendants & Bartender Helpers	410,460	HS	$18,760
35-9021	Dishwashers	502,280	Less than HS	$18,780
35-9031	Hosts and Hostesses, Restaurant, Lounge, and Coffee Shop	372,670	Less than HS	$18,720
37-2011	Janitors & Cleaners, Except Maids & Housekeeping Cleaners	2,137,730	HS	$22,840
37-2012	Maids and Housekeeping Cleaners	929,540	HS	$20,120
37-3011	Landscaping and Groundskeeping Workers	868,770	Less than HS	$24,290
39-1012	Slot Supervisors	7,000	HS	$33,270
39-2021	Nonfarm Animal Caretakers	161,820	HS	$20,340
39-3011	Gaming Dealers	96,060	HS	$18,560
39-3012	Gaming and Sports Book Writers and Runners	12,160	HS	$22,560

39-3031	Ushers, Lobby Attendants, and Ticket Takers	113,700	HS	$18,760
39-3091	Amusement and Recreation Attendants	274,230	Less than HS	$18,880
39-3092	Costume Attendants	6,270	HS	$41,670
39-3093	Locker Room, Coatroom, and Dressing Room Attendants	17,830	HS	$19,940
39-4021	Funeral Attendants	34,950	HS	$23,080
39-5011	Barbers	14,140	Post HS Certificate	$25,410
39-5012	Hairdressers, Hairstylists, and Cosmetologists	343,140	Post HS Certificate	$23,120
39-5092	Manicurists and Pedicurists	79,090	HS	$19,620
39-5093	Shampooers	16,560	Post HS Certificate	$18,760
39-5094	Skincare Specialists	38,290	Post HS Certificate	$29,050
39-6011	Baggage Porters and Bellhops	44,170	HS	$20,930
39-7011	Tour Guides and Escorts	35,100	Some College	$23,930
39-9021	Personal Care Aides	1,257,000	HS	$20,440
41-2011	Cashiers	3,398,330	HS	$19,060
41-2012	Gaming Change Persons and Booth Cashiers	19,580	HS	$23,340
41-2021	Counter and Rental Clerks	437,610	Less than HS	$23,860
41-9011	Demonstrators and Product Promoters	83,600	HS	$24,520
41-9012	Models	5,140	Less than HS	$19,970
41-9041	Telemarketers	234,520	HS	$22,740
41-9091	Door-to-Door Sales Workers, News and Street Vendors . . .	7,610	Less than HS	$21,530
43-2011	Switchboard Operators, Including Answering Service	108,890	HS	$26,550
43-2021	Telephone Operators	10,220	HS	$35,140
43-3021	Billing and Posting Clerks	490,860	HS	$34,410
43-3041	Gaming Cage Workers	16,350	HS	$25,810
43-3051	Payroll and Timekeeping Clerks	166,400	Post HS Certificate	$39,700
43-4031	Court, Municipal, and License Clerks	128,490	HS	$35,460

43-4111	Interviewers, Except Eligibility and Loan	190,710	HS	$30,790
43-4171	Receptionists and Information Clerks	981,150	HS	$26,760
43-5021	Couriers and Messengers	71,760	HS	$26,640
43-5041	Meter Readers, Utilities	36,210	HS	$37,580
43-5051	Postal Service Clerks	71,910	HS	$55,590
43-5052	Postal Service Mail Carriers	307,490	HS	$57,200
43-5053	Postal Service Mail Sorters, Processors, . . . Machine Operators	121,590	Less than HS	$54,520
43-5071	Shipping, Receiving, and Traffic Clerks	661,530	HS	$29,930
43-5081	Stock Clerks and Order Fillers	1,878,860	HS	$22,850
43-5111	Weighers, Measurers, Checkers, and Samplers, Recordkeeping	69,430	HS	$28,570
43-6012	Legal Secretaries	212,910	Post HS Certificate	$42,770
43-6013	Medical Secretaries	516,050	Post HS Certificate	$32,240
43-9021	Data Entry Keyers	205,950	HS	$28,870
43-9022	Word Processors and Typists	81,300	HS	$36,700
43-9041	Insurance Claims and Policy Processing Clerks	252,670	HS	$36,740
43-9051	Mail Clerks and Mail Machine Operators, Except Postal Service	99,190	Less than HS	$27,890
43-9061	Office Clerks, General	2,889,970	HS	$28,670
43-9081	Proofreaders and Copy Markers	10,500	BA	$34,980
45-2041	Graders and Sorters, Agricultural Products	36,100	Less than HS	$19,910
45-4023	Log Graders and Scalers	2,780	HS	$35,430
47-2042	Floor Layers, Except Carpet, Wood, and Hard Tiles	9,830	HS	$36,670
47-2043	Floor Sanders and Finishers	4,510	HS	$35,770
47-2082	Tapers	16,820	HS	$46,630
47-2121	Glaziers	42,820	HS	$38,410
47-2161	Plasterers and Stucco Masons	20,760	Less than HS	$37,550
47-3014	Helpers–Painters, Paperhangers, Plasterers, & Stucco Masons	11,570	Less than HS	$25,910
47-3015	Helpers–Pipelayers, Plumbers, Pipefitters, and Steamfitters	51,350	HS	$27,710

49-3022	Automotive Glass Installers and Repairers	15,670	HS	$32,590
49-3093	Tire Repairers and Changers	100,510	HS	$23,730
49-9064	Watch Repairers	2,390	HS	$35,450
49-9093	Fabric Menders, Except Garment	710	HS	$23,930
51-2021	Coil Winders, Tapers, and Finishers	14,930	HS	$32,980
51-2022	Electrical and Electronic Equipment Assemblers	207,330	HS	$29,910
51-2092	Team Assemblers	1,125,160	HS	$28,370
51-3011	Bakers	173,730	HS	$23,600
51-3021	Butchers and Meat Cutters	137,050	HS	$28,660
51-3022	Meat, Poultry, and Fish Cutters and Trimmers	150,310	Less than HS	$23,350
51-3023	Slaughterers and Meat Packers	86,070	Less than HS	$25,560
51-3091	Food & Tobacco Roasting, Baking, & Drying Machine Operators and Tenders	18,890	Less than HS	$27,680
51-3092	Food Batchmakers	120,850	HS	$26,770
51-3093	Food Cooking Machine Operators and Tenders	36,850	HS	$27,590
51-4052	Pourers and Casters, Metal	9,690	HS	$32,410
51-4071	Foundry Mold and Coremakers	11,870	HS	$31,340
51-4072	Molding, Coremaking, and Casting Machine Setters, Operators . . .	128,540	HS	$28,810
51-4193	Plating and Coating Machine Setters, Operators, and Tenders, Metal and Plastic	35,900	HS	$30,210
51-5111	Prepress Technicians and Workers	36,180	Post HS Certificate	$37,200
51-6011	Laundry and Dry-Cleaning Workers	199,330	HS	$20,320
51-6021	Pressers, Textile, Garment, and Related Materials	50,150	Less than HS	$20,150
51-6031	Sewing Machine Operators	142,070	Less than HS	$21,920
51-6041	Shoe and Leather Workers and Repairers	7,710	HS	$23,770
51-6051	Sewers, Hand	5,960	HS	$23,630
51-6063	Textile Knitting and Weaving Machine Setters, Operators...	22,760	HS	$27,270
51-6064	Textile Winding, Twisting, and Drawing Out Machine Setters . . .	25,740	HS	$26,250

51-7021	Furniture Finishers	15,320	Less than HS	$28,810
51-7041	Sawing Machine Setters, Operators, and Tenders, Wood	46,320	HS	$27,040
51-9031	Cutters and Trimmers, Hand	15,520	HS	$25,920
51-9051	Furnace, Kiln, Oven, Drier, and Kettle Operators and Tenders	20,590	HS	$34,900
51-9061	Inspectors, Testers, Sorters, Samplers, and Weighers	489,750	HS	$35,330
51-9111	Packaging and Filling Machine Operators and Tenders	381,760	HS	$26,410
51-9122	Painters, Transportation Equipment	49,950	HS	$40,770
51-9123	Painting, Coating, and Decorating Workers	16,280	Less than HS	$28,750
51-9192	Cleaning, Washing, and Metal Pickling Equipment Operators and Tenders	17,360	HS	$26,910
51-9197	Tire Builders	17,680	HS	$42,540
51-9198	Helpers–Production Workers	420,520	HS	$23,610
53-3021	Bus Drivers, Transit and Intercity	158,050	HS	$37,470
53-3022	Bus Drivers, School or Special Client	499,440	HS	$28,850
53-3031	Driver/Sales Workers	405,810	HS	$22,250
53-3033	Light Truck or Delivery Services Drivers	797,010	HS	$29,570
53-3041	Taxi Drivers and Chauffeurs	178,260	HS	$23,210
53-4021	Railroad Brake, Signal, and Switch Operators	21,060	HS	$52,360
53-6021	Parking Lot Attendants	136,440	HS	$19,800
53-6031	Automotive and Watercraft Service Attendants	104,750	HS	$20,900
53-7041	Hoist and Winch Operators	2,840	HS	$39,580
53-7051	Industrial Truck and Tractor Operators	521,840	HS	$31,340
53-7062	Laborers and Freight, Stock, and Material Movers, Hand	2,400,490	HS	$24,430
53-7063	Machine Feeders and Offbearers	104,340	HS	$29,290
53-7064	Packers and Packagers, Hand	693,170	HS	$20,330
53-7081	Refuse and Recyclable Material Collectors	115,170	HS	$33,660
53-7111	Mine Shuttle Car Operators	2,630	HS	$55,000

SOURCE: O*NET and OES 2014; author calculations.

References

Abel, Jaison, and Richard Deitz. 2016. "Underemployment in the Early Careers of College Graduates Following the Great Recession." Staff Report No. 749. New York: Federal Reserve Bank of New York. https://www.newyorkfed.org/medialibrary/media/research/staff_reports/sr749.pdf?la=en (accessed April 11, 2017).

Abel, Jaison R., and Todd M. Gabe. 2008. "Human Capital and Economic Activity in Urban America." Staff Report No. 332. New York: Federal Reserve Bank of New York.

Acemoglu, Daron. 1998. "Why Do New Technologies Complement Skills? Directed Technical Change and Wage Inequality." *Quarterly Journal of Economics* 113(4): 1055–1089.

American Community Survey. 2013. 5-Year Estimate. Suitland, MD: U.S. Census Bureau. https://www.census.gov/programs-surveys/acs/ (accessed April 11, 2017).

Andreason, Stuart. 2015. "Will Talent Attraction and Retention Improve Metropolitan Labor Markets? The Labor Market Impact of Increased Educational Attainment in U.S. Metropolitan Regions, 1990–2010." Working Paper No. 2015-4. Atlanta: Federal Reserve Bank of Atlanta.

Arrow, Kenneth. 1962. "The Economic Implications of Learning by Doing." *Review of Economic Studies* 29(3): 155–173.

Autor, David. 2010. "The Polarization of Job Opportunities in the U.S. Labor Market: Implications for Employment and Earnings." Washington, DC: Center for American Progress and The Hamilton Project. https://economics.mit.edu/files/5554 (accessed June 23, 2017).

Autor, David H., Lawrence F. Katz, and Melissa Kearney. 2008. "Trends in U.S. Wage Inequality: Revising the Revisionists." *Review of Economics and Statistics* 90(2): 300–323.

Autor, David H., Frank Levy, and Richard J. Murnane. 2002. "Upstairs, Downstairs: Computers and Skills on Two Floors of a Large Bank." *Industrial and Labor Relations Review* 55(3): 432–447.

———. 2003. "The Skill Content of Recent Technological Change: An Empirical Exploration." *Quarterly Journal of Economics* 118(4): 1279–1333.

Bacolod, Marigee, Bernard S. Blum, and William Strange. 2010. "Elements of Skill: Traits, Intelligences, Education, and Agglomeration." *Journal of Regional Science* 50(10): 245–280.

Barney, Jay. 1991. "Firm Resources and Sustained Competitive Advantage." *Journal of Management* 17(1): 99–120.

Bartik, Timothy J. 1992. *Who Benefits from State and Local Economic Devel-*

opment Policies? Kalamazoo, MI: W.E. Upjohn Institute for Employment Research.

———. 1993. "Who Benefits from Local Job Growth: Migrants or the Original Residents?" *Regional Studies* 27(4): 297–311.

———. 2011. *Investing in Kids: Early Childhood Programs and Local Economic Development.* Kalamazoo, MI: W.E. Upjohn Institute for Employment Research.

Baum, Sandy, and Jennifer Ma. 2007. *Education Pays 2007: The Benefits of Higher Education for Individuals and Society.* Washington, DC: College Board.

Beam, Adam. 2016. "Kentucky Gov. Matt Bevin Wants State Colleges and Universities to Produce More Electrical Engineers and Less French Literature Scholars." Associated Press, January 29. http://www.usnews .com/news/us/articles/2016-01-29/in-kentucky-a-push-for-engineers-over -french-lit-scholars (accessed April 11, 2017).

Becker, Gary. 1962. "Investment in Human Capital: A Theoretical Analysis." *Journal of Political Economy* 70(5): 9–49.

———. 1964/1993. *Human Capital: A Theoretical and Empirical Analysis with Special Reference to Education.* 3rd ed. Chicago: University of Chicago Press.

Benhabib, Jess, and Mark M. Spiegel. 1994. "The Role of Human Capital in Economic Development: Evidence from Aggregate Cross-Country Data." *Journal of Monetary Economics* 34(2): 143–173.

Black, Duncan, and Vernon Henderson. 1999. "A Theory of Urban Growth." *Journal of Political Economy* 107(2): 252–284.

Blumenthal, Pamela, Harold L. Wolman, and Edward W. Hill. 2009. "Understanding the Economic Performance of Metropolitan Areas in the United States." *Urban Studies* 46(3): 605–627.

Borghans, Lex, Bas ter Weel, and Bruce A. Weinberg. 2014. "People Skills and the Labor-Market Outcomes of Underrepresented Groups." *Industrial and Labor Relations Review* 67(2): 287–334.

Breuninger, Kevin. 2017. "Trump Wants 4.5 Million New Apprenticeships in Five Years—With Nearly the Same Budget." CNBC, June 15. https://www .cnbc.com/2017/06/15/trump-wants-a-10-fold-increase-in-apprenticeships -in-five-years-with-the-same-budget.html (accessed September 20, 2017).

Brink, Kyle E., and Robert D. Costigan. 2015. "Oral Communication Skills: Are the Priorities of the Workplace and AACSB-Accredited Business Programs Aligned?" *Academy of Management Learning and Education* 14(2): 205–221.

Brynjolfsson, Erik, Lorin Hitt, and Heekyung Kim. 2011. "Strength in Numbers: How Does Data-Driven Decisionmaking Affect Firm Performance?"

Working paper. Cambridge, MA: The MIT Center for Digital Business. http://www.ebusiness.mit.edu/research/papers/2011.12_Brynjolfsson_Hitt _kim_Strength in Numbers_302.pdf (accessed April 12, 2017).

Bui, Quoctrung. 2016. "The States that College Graduates are Most Likely to Leave." *New York Times* Upshot, November 22. https://www.nytimes .com/2016/11/22/upshot/the-states-that-college-graduates-are-most-likely -to-leave.html (accessed September 22, 2017).

Bureau of Labor Statistics. n.d.-a. Occupational Employment Statistics. Washington, DC: Bureau of Labor Statistics. http://www.bls.gov/oes/ accessed April 12, 2017).

———. n.d.-b. Education and Training Outlook for Occupations, 2012–2022. Washington, DC: Bureau of Labor Statistics. http://www.dallastown.net/ cms/lib6/PA01000011/Centricity/Domain/97/ep_edtrain_outlook.pdf (accessed June 21, 2017).

Capozza, Dennis R, Patric H. Hendershott, Charlotte Mack, and Christopher J. Mayer. 2002. "Determinants of Real House Price Dynamics." NBER Working Paper No. 9262. Cambridge, MA: National Bureau of Economic Research.

Cappelli, Peter. 2012. "The Skills Gap Myth: Why Companies Can't Find Good People." *Time*, June 4. http://business.time.com/2012/06/04/the-skills-gap -myth-why-companies-cant-find-good-people/ (accessed June 23, 2017).

———. 2015. "Skill Gaps, Skill Shortages, and Skill Mismatches: Evidence and Arguments for the United States." *Industrial and Labor Relations Review* 68(2): 251–290.

Carnevale, Anthony P., Ben Cheah, and Andrew R. Hanson. 2015. *The Economic Value of College Majors*. Washington, DC: Center on Education and the Workforce, Georgetown University. https://www.luminafoundation.org/ files/resources/economic-value-of-college-majors.pdf (accessed April 12, 2017).

Carnevale, Anthony P., Tamara Jayasundera, and Artem Gulish. 2016. *America's Divided Recovery: College Haves and Have Nots*. Washington, DC: Center on Education and the Workforce, Georgetown University. https:// cew.georgetown.edu/wp-content/uploads/Americas-Divided-Recovery -web.pdf (accessed June 23, 2017).

Carnevale, Anthony P., Nicole Smith, and Michelle Melton. 2011. *STEM*. Washington, DC: Center on Education and the Workforce, Georgetown University. https://cew.georgetown.edu/wp-content/uploads/2014/11/stem -complete.pdf (accessed June 23, 2017).

Carnevale, Anthony P., Nicole Smith, and Jeff Strohl. 2010. *Help Wanted: Projections of Job and Educational Requirements through 2018*. Washington, DC: Center on Education and the Workforce, Georgetown University.

Caselli, Francesco, Gerardo Esquivel, and Fernando Lefort. 1996. "Reopening the Convergence Debate: A New Look at Cross-Country Growth Empirics." *Journal of Economic Growth* 1(3): 363–389.

Chrisinger, Colleen K., Christopher S. Fowler, and Rachel Garshick Kleit. 2012. "Shared Skills: Occupation Clusters for Poverty Alleviation and Economic Development in the U.S." *Urban Studies* 49(15): 3403–3425.

Combes, Pierre-Phillippe, Gille Duranton, Laurent Gobillon, Diego Puga, and Sébastien Roux. 2012. "The Productivity Advantages of Large Cities: Distinguishing Agglomeration from Firm Selection." *Econometrica* 80(6): 2543–2594.

Cooper, Arnold C., F. Javier Gimeno-Gascon, and Carolyn Y. Woo. 1994. "Initial Human and Financial Capital as Predictors of New Venture Performance." *Journal of Business Venturing* 9(5): 371–395.

Davidson, Adam. 2012. "Skills Don't Pay the Bills." *New York Times Magazine*, November 20. http://www.nytimes.com/2012/11/25/magazine/skills -dont-pay-the-bills.html (accessed June 23, 2017).

Dill, Kathryn. 2015. "The 10 Most Innovative Tech Hubs in the U.S." *Forbes*, February 12. http://www.forbes.com/sites/kathryndill/2015/02/12/the-10 -most-innovative-tech-hubs-in-the-u-s/#72368a1c1386 (accessed April 12, 2017).

Drucker, Peter F. (1969/)1992. *Age of Discontinuity: Guidelines to Our Changing Society*. London: Transaction.

Ehrlich, Isaac. 2007. "The Mystery of Human Capital as Engine of Growth, Or Why the U.S. Became the Economic Superpower in the 20th Century." NBER Working Paper No. 12868. Cambridge, MA: National Bureau of Economic Research.

Ellis, William E. 2011. *A History of Education in Kentucky*. Lexington, KY: University of Kentucky Press.

Elvery, Joel. 2010. "City Size and Skill Intensity." *Regional Science and Urban Economics* 40(6): 367–379.

Feser, Edward J. 2003. "What Regions Do Rather than Make: A Proposed Set of Knowledge-Based Occupation Clusters." *Urban Studies* 40(10): 1937–1958.

Feser, Edward J., and Edward M. Bergman. 2000. "National Industry Cluster Templates: A Framework for Applied Regional Cluster Analysis." *Regional Studies* 34(1): 1–19.

Flamm, Kenneth. 1988. *Creating the Computer: Government, Industry and High Technology*. Washington, DC: Brookings Institution.

Florida, Richard, Charlotte Mellander, Kevin Stolarick, and Adrienne Ross. 2012. "Cities, Skills and Wages." *Journal of Economic Geography* 12(2): 355–377.

Friedman, Milton. 1955. "The Role of Government in Education." In *Economics and the Public Interest*, Robert A. Solo, ed. New Brunswick, NJ: Rutgers University Press, pp. 127–134. http://la.utexas.edu/users/hcleaver/ 330T/350kPEEFriedmanRoleOfGovttable.pdf (accessed June 23, 2017).

Gibbons, Robert, and Michael Waldman. 2004. "Task-Specific Human Capital." *AEA Papers and Proceedings* 94(2): 203–207.

Glaeser, Edward L., and David C. Mare. 2001. "Cities and Skills." *Journal of Labor Economics* 19(2): 316–342.

Glaeser, Edward L., and Matthew G. Resseger. 2010. "The Complementarity between Cities and Skills." *Journal of Regional Science* 50(1): 221–244.

Glaeser, Edward L., and Albert Saiz. 2003. "The Rise of the Skilled City." NBER Working Paper No. 10191. Cambridge, MA: National Bureau of Economic Research.

Goldin, Claudia, and Lawrence F. Katz. 2010. *The Race between Education and Technology*. Cambridge, MA: Belknap.

Goos, Maarten, and Alan Manning. 2007. "Lousy and Lovely Jobs: The Rising Polarization of Work in Britain." *Review of Economics and Statistics* 89(1): 118–133.

Gottlieb, Paul D., and Michael Fogarty. 2003. "Educational Attainment and Metropolitan Growth." *Economic Development Quarterly* 17(4): 325–336.

Gould, Eric D. 2007. "Cities, Workers, and Wages: A Structural Analysis of the Urban Wage Premium." *The Review of Economic Studies* 74(2): 477–506.

Greenough, John. 2016. "10 Million Self-Driving Cars Will Be On the Road by 2020." *Business Insider*. http://www.businessinsider.com/report-10 -million-self-driving-cars-will-be-on-the-road-by-2020-2015-5-6#comments (accessed April 12, 2017).

Hadden, Wilbur C., Nataliya Kravets, and Carles Muntaner. 2004. "Descriptive Dimensions of U.S. Occupations with Data from the O*NET." *Social Science Research* 33(1): 64–78.

Holzer, Harry J. 2008. "Workforce Development as an Antipoverty Strategy: What Do We Know? What Should We Do?" IZA Discussion Paper No. 3776: Bonn: Institute of Labor Economics.

———. 2015. "Job Market Polarization and U.S. Worker Skills: A Tale of Two Middles." Washington, DC: The Brookings Institution. https://www .brookings.edu/wp-content/uploads/2016/06/polarization_jobs_policy _holzer.pdf (accessed April 12, 2017).

Holzer, Harry J., and Robert I. Lerman. 2007. "America's Forgotten Middle-Skill Jobs: Education and Training Requirements in the Next Decade and Beyond." Washington, DC: The Workforce Alliance.

———. 2009. "The Future of Middle-Skill Jobs." CCF Brief No. 41. Washington, DC: Center on Children and Families, The Brookings Institution.

https://www.brookings.edu/wp-ontent/uploads/2016/06/02_middle_skill _jobs_holzer.pdf (accessed June 23, 2017).

Islam, Nazrul. 1995. "Growth Empirics: A Panel Data Approach." *Quarterly Journal of Economics* 110(4): 1127–1170.

Jaimovich, Nir, and Henry E. Siu. 2012. "The Trend Is the Cycle: Job Polarization and Jobless Recoveries." NBER Working Paper No. 18334. Cambridge, MA: National Bureau of Economic Research.

Kodrzycki, Yolanda K., and Ana Patricia Muñoz. 2013. "Economic Distress and Resurgence in U.S. Central Cities: Concepts, Causes, and Policy Levers." Public Policy Discussion Paper No. 13-3. Boston: Federal Reserve Bank of Boston.

Koo, Jun. 2005. "How to Analyze Regional Economy with Occupation Data." *Economic Development Quarterly* 19(4): 356–372.

Kroeger, Teresa, Tanyelle Cooke, and Elise Gould. 2016. "The Class of 2016: The Labor Market Is Still Far from Ideal for Young Graduates." Report. Washington, DC: Economic Policy Institute. http://www.epi.org/publication/ class-of-2016/ (accessed April 12, 2017).

Lerman, Robert I. 2008. "Are Skills the Problem? Reforming the Education and Training System in the United States." In *A Future of Good Jobs? America's Challenge in the Global Economy*, Timothy J. Bartik and Susan N. Houseman, eds. Kalamazoo, MI: W.E. Upjohn Institute for Employment Research, pp. 17–80.

Lucas, Robert E. 1988. "On the Mechanics of Economic Development." *Journal of Monetary Economics* 22(1): 3–42.

———. 2009. "Ideas and Growth." *Economica* 76(301): 1–19.

Markusen, Ann R. 1984. *Defense Spending and the Geography of High Tech Industries*. Berkeley: University of California, Berkeley: Institute of Urban and Regional Development.

Markusen, Ann R., Gregory H. Wassall, Douglas DeNatale, and Randy Cohen. 2008. "Defining the Creative Economy: Industry and Occupational Approaches." *Economic Development Quarterly* 22(1): 24–45.

Martin, Ron, and Peter Sunley. 2006. "Path Dependence and Regional Economic Evolution." *Journal of Economic Geography* 6(4): 395–437.

Maxwell, Nan L. 2008. "Wage Differentials, Skills, and Institutions in Low-Skill Jobs." *Industrial and Labor Relations Review* 61(3): 394–409.

Moretti, Enrico. 2004. "Workers' Education, Spillovers, and Productivity: Evidence from Plant-Level Production Functions." *American Economic Review* 94(3): 656–690.

National Center for Education Statistics. n.d. Career and Technical Education Statistics, Table H126. Percentage of Public High School Graduates Who Earned Credits, by Curricular Area: 1990, 2000, 2005, and 2009. Washington, DC: NCES. https://nces.ed.gov/surveys/ctes/tables/h126.asp (accessed June 23, 2017).

National Conference of State Legislatures. 2015. *Performance-Based Funding for Higher Education*. Denver, CO: NCSL. http://www.ncsl.org/research/education/performance-funding.aspx (accessed April 12, 2017).

National Education Association. 2012. *Reality Check: The U.S. Job Market and Students' Academic and Career Paths Necessitate Enhanced Vocational Education in High Schools*. Washington, DC: NEA. http://www.nea.org/assets/docs/Vocational_Education_final.pdf (accessed June 23, 2017).

National Science Board. 2015. *Revisiting the STEM Workforce: A Companion to Science and Engineering Indicators 2014*. Arlington, VA: NSB. https://www.nsf.gov/nsb/publications/2015/nsb201510.pdf (accessed June 23, 2017).

National Skills Coalition. 2017. *United States' Forgotten Middle*. Washington, DC: National Skills Coalition. http://www.nationalskillscoalition.org/resources/publications/2017-middle-skills-fact-sheets/file/United-States-MiddleSkills.pdf (accessed June 23, 2017).

Nelson, Richard R., and Edmund S. Phelps. 1966. "Investment in Humans, Technological Diffusion, and Economic Growth." *American Economic Review* 56(1/2): 69–75.

North, Douglass C. 1990. *Institutions, Institutional Change and Economic Performance*. New York: Cambridge University Press.

Office of the President. 2012. *Preparing a 21st Century Workforce: Science, Technology, Engineering, and Mathematics (STEM) Education in the 2014 Budget*. Washington, DC: Office of the President. https://obamawhitehouse.archives.gov/sites/default/files/microsites/ostp/2014_Ran&Dbudget_STEM.pdf (accessed April 12, 2017).

Osterman, Paul, and Andrew Weaver. 2014. "Why Claims of Skills Shortages in Manufacturing Are Overblown." Washington, DC: Economic Policy Institute. http://www.epi.org/publication/claims-skills-shortages-manufacturing-overblown/ (accessed April 12, 2017).

Papka, Perry. 2017. "A New Postsecondary Funding Model: A Positive Step for Future Investments, but Details and Transparency Matter." The Prichard Blog, February 24. Lexington, KY: Prichard Committee for Academic Excellence. http://prichblog.blogspot.com/2017/02/a-new-postsecondary-funding-model.html (accessed June 27, 2017).

Partridge, Mark D., and Dan S. Rickman. 2003. "The Waxing and Waning of Regional Economies: The Chicken-Egg Question of Jobs vs. People." *Journal of Urban Economics* 53(1): 76–97.

Rauch, James E. 1993. "Productivity Gains from Geographic Concentration of Human Capital: Evidence from the Cities." *Journal of Urban Economics* 34(3): 380–400.

Robles, Marcel M. 2012. "Executive Perceptions of the Top 10 Soft Skills Needed in Today's Workplace." *Business Communications Quarterly* 75(4): 453–465.

Romer, Paul M. 1990. "Endogenous Technological Change." *Journal of Political Economy* 98(5): S71–S102.

Ross, Martha, and Nicole Prchal Svajlenka. 2016. *Employment and Disconnection among Teens and Young Adults: The Role of Place, Race, and Education.* Report. Washington, DC: The Brookings Institution. http://www .brookings.edu/research/reports2/2016/05/24-teen-young-adult -employment-recession-ross-svajlenka#V0G0 (accessed April 12, 2017).

Rothwell, Jonathan. 2013. *The Hidden STEM Economy.* Report. Washington, DC: Metropolitan Policy Program, The Brookings Institution. https://www .brookings.edu/research/the-hidden-stem-economy/ (accessed April 12, 2017).

Ruckelshaus, Catherine, and Sarah Leberstein. 2014. *Manufacturing Low Pay: Declining Wages in the Jobs That Built America's Middle Class.* New York: National Employment Law Project. http://www.nelp.org/content/ uploads/2015/03/Manufacturing-Low-Pay-Declining-Wages-Jobs-Built -Middle-Class.pdf (accessed June 23, 2017).

Schultz, Theodore W. 1961. "Investment in Human Capital." *American Economic Review* 51(1): 1–17.

Scott, Allen J. 2009. "Human Capital Resources and Requirements across the Metropolitan Hierarchy of the USA." *Journal of Economic Geography* 9(2): 207–226.

Scott, Allen J., and Agostino Mantegna. 2009. "Human Capital Assets and Structures of Work in the U.S. Metropolitan Hierarchy (An Analysis Based on the O*NET Information System)." *International Regional Science Review* 32(2): 173–194.

Shapiro, Jesse M. 2006. "Smart Cities: Quality of Life, Productivity, and the Growth Effects of Human Capital." *Review of Economics and Statistics* 88(2): 324–335.

Simon, Curtis J. 1998. "Human Capital and Metropolitan Employment Growth." *Journal of Urban Economics* 43(2): 223–243.

Simon, Curtis J., and Clark Nardinelli. 2002. "Human Capital and the Rise of American Cities, 1900–1990." *Regional Science and Urban Economics* 32(1): 59–96.

Smith, Adam. (1776/)2008. *An Inquiry into the Nature and Causes of the Wealth of Nations.* Oxford: Oxford's World Classics.

Teitelbaum, Michael. 2014. *Falling Behind? Boom, Bust and the Global Race for Scientific Talent.* Princeton, NJ: Princeton University Press.

Torpey, Elka. 2013. "College to Career: Projected Job Openings in Occupations That Typically Require a Bachelor's Degree." *Occupations Outlook Quarterly* (Summer). https://www.bls.gov/careeroutlook/2013/summer/art03.pdf (accessed June 23, 2017).

University of Illinois Department of Computer Science. n.d.-a. "History Timeline." Urbana: University of Illinois. https://cs.illinois.edu/about-us/history-timeline (accessed June 23, 2017).

———. n.d.-b. "Statistics." Urbana: University of Illinois. https://cs.illinois.edu/about-us/statistics (accessed June 23, 2017).

U.S. Department of Labor (USDOL). Employment and Training Administration. n.d. O*NET. Washington, DC: USDOL. http://www.onetonline.org (accessed September 21, 2017).

Vilorio, D. 2014. "STEM 101: Intro to Tomorrow's Jobs." *Occupational Outlook Quarterly* (Spring). https://www.bls.gov/careeroutlook/2014/spring/art01.pdf (accessed June 21, 2017).

Weber, Lauren. 2014. "Apprenticeships Help Close the Skills Gap. So Why Are They in Decline? Some States Try Extending the Practice to More Professions." *Wall Street Journal*, April 27. http://www.wsj.com/news/articles/SB10001424052702303978304579473501943642612 (accessed April 12, 2017).

Wolfe, David A., and Meric S. Gertler 2004. "Clusters from the Inside and Out: Local Dynamics and Global Linkages." *Urban Studies* 41(5/6): 1071–1093.

Wolf-Powers, Laura. 2013. "Predictors of Employment Growth and Unemployment in U.S. Central Cities, 1990–2010." Upjohn Institute Working Paper 13-199. Kalamazoo, MI: W.E. Upjohn Institute for Employment Research.

Xue, Yi, and Richard C. Larson. 2015. "STEM Crisis or STEM Surplus? Yes and Yes." *Monthly Labor Review* (May). https://www.bls.gov/opub/mlr/2015/article/stem-crisis-or-stem-surplus-yes-and-yes.htm (accessed June 23, 2017).

Yakusheva, Olga. 2010. "Return to College Education Revisited: Is Relevance Relevant?" *Economics of Education Review* 29(6): 1125–1142.

Author

Fran Stewart is a Cleveland-area independent researcher and writer. She has spent two decades developing and contributing to public-policy reports on economic development and urban issues. Her body of work comprises projects for state government agencies, universities, nonprofit economic development intermediaries, and local think tanks and advocacy groups. Her research interests include skill development, regional economic resilience, economic driver industries, and manufacturing. Stewart is the senior research fellow for the Ohio Manufacturing Institute, a research and advocacy center at the Ohio State University's College of Engineering that explores public-policy issues related to manufacturing and economic development. A former newspaper journalist, Stewart has written extensively on topics ranging from education, particularly higher education, and immigration to neighborhoods, and tourism. Over the years, the challenge of preparing young people for a rapidly changing work environment has become a driving interest. Stewart holds a PhD in urban studies from Cleveland State University.

Index

Note: The italic letters *f, n,* or *t* following a page number indicate a figure, note or table, respectively, on that page. Double letters mean more than one such consecutive item on a single page.

About the Institute

The W.E. Upjohn Institute for Employment Research is a nonprofit research organization devoted to finding and promoting solutions to employment-related problems at the national, state, and local levels. It is an activity of the W.E. Upjohn Unemployment Trustee Corporation, which was established in 1932 to administer a fund set aside by Dr. W.E. Upjohn, founder of The Upjohn Company, to seek ways to counteract the loss of employment income during economic downturns.

The Institute is funded largely by income from the W.E. Upjohn Unemployment Trust, supplemented by outside grants, contracts, and sales of publications. Activities of the Institute comprise the following elements: 1) a research program conducted by a resident staff of professional social scientists; 2) a competitive grant program, which expands and complements the internal research program by providing financial support to researchers outside the Institute; 3) a publications program, which provides the major vehicle for disseminating the research of staff and grantees, as well as other selected works in the field; and 4) an Employment Management Services division, which manages most of the publicly funded employment and training programs in the local area.

The broad objectives of the Institute's research, grant, and publication programs are to 1) promote scholarship and experimentation on issues of public and private employment and unemployment policy, and 2) make knowledge and scholarship relevant and useful to policymakers in their pursuit of solutions to employment and unemployment problems.

Current areas of concentration for these programs include causes, consequences, and measures to alleviate unemployment; social insurance and income maintenance programs; compensation; workforce quality; work arrangements; family labor issues; labor-management relations; and regional economic development and local labor markets.